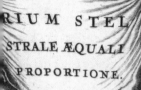

THE ART OF THE
MAP

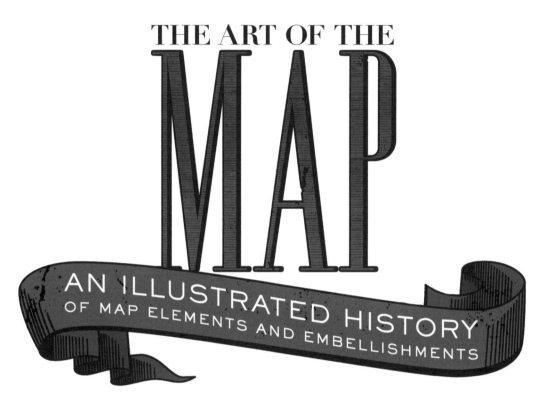

AN ILLUSTRATED HISTORY
OF MAP ELEMENTS AND EMBELLISHMENTS

DENNIS REINHARTZ

FOREWORD BY
JOHN NOBLE WILFORD

STERLING
New York

STERLING
New York

An Imprint of Sterling Publishing
387 Park Avenue South
New York, NY 10016

© 2012 by Dennis Reinhartz
For illustration credits see page 218

Book design and production: gonzalez defino, ny / gonzalezdefino.com

ISBN 978-1-4027-6592-6

Distributed in Canada by Sterling Publishing
c/o Canadian Manda Group, 165 Dufferin Street
Toronto, Ontario, Canada M6K 3H6
Distributed in the United Kingdom by GMC Distribution Services
Castle Place, 166 High Street, Lewes, East Sussex, England BN7 1XU
Distributed in Australia by Capricorn Link (Australia) Pty. Ltd.
P.O. Box 704, Windsor, NSW 2756, Australia

Courtesy of Geography & Map Division, Library of Congress:
FRONTISPIECE—Andreas Cellarius, *Harmonia Macrocosmica* . . . (detail), 1708, g319mgct00059;
page v: Abraham Ortelius, *Theatrum Orbis Terrarum* (title page), 1570, g3200m gct00003.

For information about custom editions, special sales, and premium and corporate purchases,
please contact Sterling Special Sales at 800-805-5489 or specialsales@sterlingpublishing.com.

Manufactured in Canada

2 4 6 8 10 9 7 5 3 1

www.sterlingpublishing.com

To Friends and mentors
Eila, Helen, Brian,
and David
And, as always,
to Judy

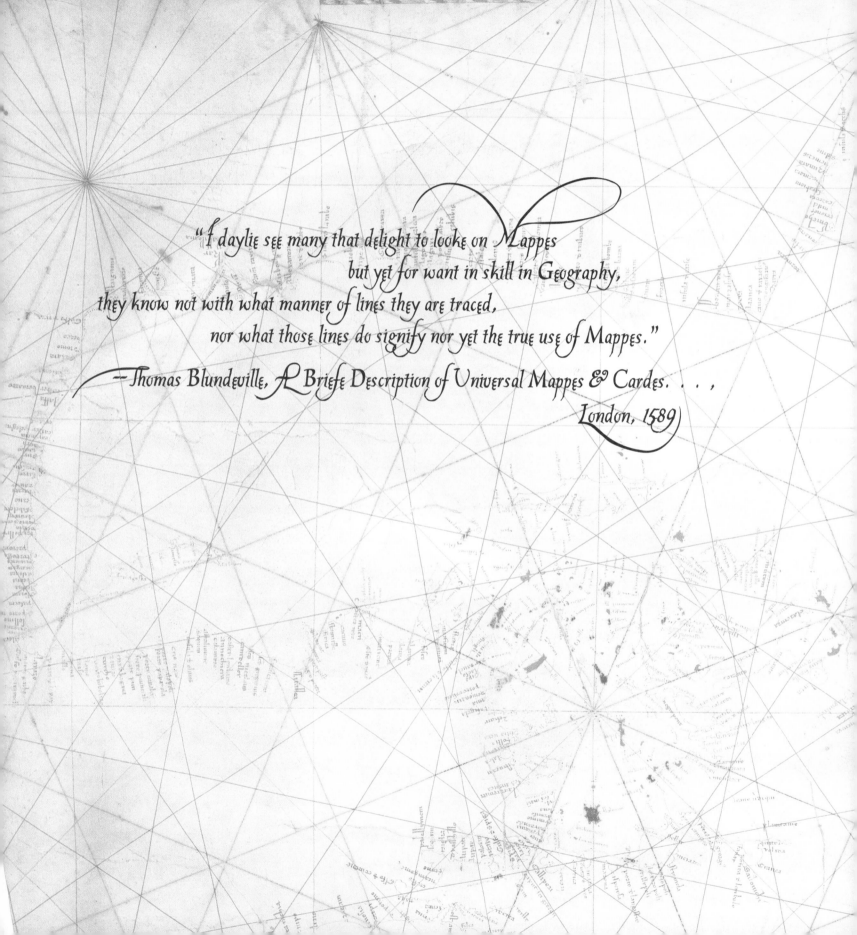

"I daylie see many that delight to looke on Mappes
but yet for want in skill in Geography,
they know not with what manner of lines they are traced,
nor what those lines do signify nor yet the true use of Mappes."

—Thomas Blundeville, A Briefe Description of Universal Mappes & Cardes. . . ,
London, 1589

Contents

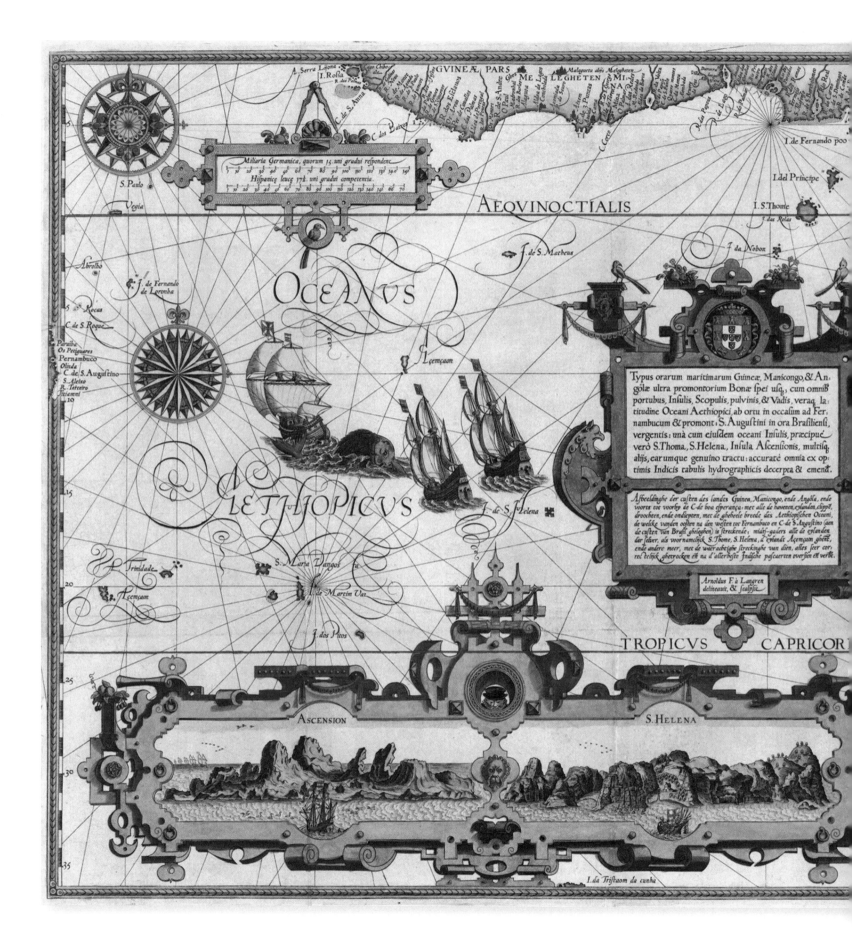

ho has not unfolded, unrolled, or otherwise spread out a map and felt all its promise of journeys ahead, planned or only imagined? Who has not asked a map to point the way? To the other side of the river, or the mountain. To the shore, or lands beyond borders and deep waters. Who has not learned from a map one's place in the world, some sense of *here* in relation to *there*?

Joseph Conrad understood the feeling behind such questions. In *Heart of Darkness*, Conrad has Marlowe say: "Now when I was a little chap I had a passion for maps. I would look for hours at South America, or Africa, or Australia, and lose myself in all the glories of exploration. At that time there were many blank spaces on the earth, and when I saw one that looked particularly inviting on a map (but they all look that) I would put my finger on it and say, 'When I grow up I will go there.'"

Many of us no doubt can single out a moment of discovering the fascination of maps, their power and indispensability. I recall that I was a little chap of eight, in the third grade, when I made the discovery. It was a momentous date, December 7, 1941. To President Franklin D. Roosevelt, it was "a date which will live in infamy." This surely means war for us, I heard my parents say, and radio newscasters agreed with equal gravity. For me, the surprise bombing attack by Japanese aircraft on Pearl Harbor raised for the first time urgent geographical questions that only a map could answer with graphic efficiency. I had never heard of Pearl Harbor and not much about Hawaii, except that pineapples grow there. We were then a country of forty-eight states, confined to the North American continent. If I had been in the third grade in 1914, I suppose I would have been just as bewildered over the news of a Hapsburg archduke's assassination at Sarajevo.

Jan Huygen van Linschoten, Typus Orarum Maritimarum Guineae . . . , *Amsterdam, 1596.*

Anyway, as this was before television, much less the Internet, the first news reached us without informative maps. We had no adequate atlas at home, only gas-station road maps (of Kentucky, Tennessee, and nearby states) and a jigsaw puzzle with pieces in the shape of the individual states, by which I had learned their names and capitals. My mother or father suggested that I look up Hawaii in the encyclopedia. The entry, though skimpy, did include a small map of the islands, with an inset showing their relation to the American West Coast. In other volumes, maps showed the Pacific Ocean. It was a relief to see how far away Japan seemed, even Hawaii.

We eventually got a world map, and I faithfully clipped maps of military operations out of newspapers and magazines. Through war and maps, well before I might normally have, I was learning world geography. I know that reading and listening to war news influenced my choice of a career in journalism. And I suspect this also was the source of my long interest in maps and the inspiration to write *The Mapmakers*, a history of cartography.

We are told that no one knows when or where or for what purpose someone got the first idea to draw a map. It was presumably thousands of years ago, before written language, and it most likely occurred independently among different peoples in separate parts of the world. All the great ancient civilizations—Egypt, Mesopotamia, and China—evolved the map idea at early stages. The earliest known world map, after a fashion, is a Babylonian clay tablet from the sixth century BCE. Long before that, prehistoric Europeans drew sketch maps on their cave walls. And before European contact, the

Incas made elaborate relief maps of stone and clay and Pacific islanders prepared maps of sticks lashed together with fibers to depict prevailing wind and wave patterns, with shells or coral inserted to represent islands.

Writing in *The History of Cartography*, an ongoing, authoritative multivolume work (1987–present), one of its editors, Brian Harley, observed, "There has probably always been a mapping impulse in human consciousness, and the mapping experience—involving the cognitive mapping of space—undoubtedly existed long before the physical artifacts we now call maps."

Judging by reactions to my book, map enthusiasts are many and varied. Maps are sources of pleasure as guides to travel and idle dreaming, and objects of aesthetic value among collectors. They are instruments of exploration and navigation in the air and outer space, on land, and on sea. They both inspire and illustrate scientific insights, as in the drifting continents and seafloor spreading of plate tectonics. They are a metaphor in the hands of generals and admirals "mapping their campaigns." Historians, as Dennis Reinhartz notes in the preface to this book, look to maps to "show where historical events unfolded and how those places were perceived during past eras" and help explain "why events in history happened as they did, where they did."

Dr. Reinhartz, a historian of cartography, has been working with maps for more than forty years. In this book, he concentrates on maps as informative art forms, emphasizing the most important visual elements of cartography used by Europeans in the fifteenth to the nineteenth centuries. It is an inspired approach to the story of the expanding content and uses of maps produced in what are known as the Age of Discovery and the Enlightenment, a time of spreading empires. Maps both recorded and encouraged such enterprises. "Traditionally," the author writes, "there has been a symbiosis between discovery, exploration, exploitation of resources, and cartography."

As aids to historical understanding, the maps of this period illustrate the steady progress in geographical knowledge, which is to say the gradual filling in of the blank spaces on previous maps. This was the time when mapmaking emerged from the cloistered scriptoria and, with the coming of the printing press, became a more robust enterprise enjoying royal patronage. Early on, following the introduction of the compass in Europe, the making and selling of portolan charts for Mediterranean mariners may have first taken cartography into commercial waters.

A special delight of Dr. Reinhartz's book is his discussion of the little fancies and flourishes that decorate old maps. It seems that every mapmaker back then was a repressed artist who couldn't wait to be done with shorelines and rhumb lines. When it came to orienting the map, the inner artist felt free to embellish the necessary with symbolic blossoms—compass roses—spreading in the cardinal directions. In other flights of whimsy, cherubs with chubby cheeks blow in the directions of the prevailing winds. These features drive up old-map prices at auction.

Modern maps are more comprehensive, reliable, and down-to-business dependable than their predecessors. As a practical matter, we can be thankful. But this progress robs us of the charms of old maps and charts, as satirized by Jonathan Swift in "On Poetry, A Rhapsody" (1733): "Geographers, in Afric maps/With savage pictures fill their gaps."

Early mapmakers employed just about every artifice imaginable to fill their gaps in geographical knowledge. On uncharted expanses of land, elephants were indeed drawn on seventeenth-century maps of Africa, and beavers were depicted on many British and French maps of the North American hinterland. Mapmakers met the challenge of the open seas with drawings of ships in full sail, whales and flying fishes, nude mermaids, and imaginary islands. As an aside, Dr. Reinhartz tells us that dragon-like creatures appeared on many maps, but not, as generally believed, accompanied by the Latin phrase meaning "Here be dragons." These words appear on only one small copper globe from the early sixteenth century.

Blank spaces on maps have significance for those of us who call ourselves Americans. We owe the nomenclature of our hemisphere to the vacant interior on a 1507 map of the New World by Martin Waldseemüller. On what is now Brazil, the mapmaker inscribed the name "America," in the mistaken belief that Amerigo Vespucci, not Columbus, deserved credit for first sighting a part of the continent of South America. On a revised map, Waldseemüller withdrew the name, but it was too late. The only known original copy of the map that named us resides in the Library of Congress, a tribute to the power of maps.

—John Noble Wilford

יְהֹוָה

A New & Correct MAP of the

WHOLE WORLD

Shewing y̆ Situation of its Principal Parts. Viz the
Oceans, Kingdoms, Rivers, Capes, Ports, Moun-
tains, Woods. Trade-Winds, Monsoons, Variation of y̆
Compass, Climats, &c. With the most Remarkable
Tracks of the Bold Attempts which have been made to
Find out the North East & North West Passages.

The Projection of this Map is Call'd Mercator's the Design is
to make it Useful both for Land and Sea. And it is laid
Down with all possible Care, According to the Newest and
Most Exact Observations By

HERMAN MOLL Geographer.

1719

L. Cheron delin.

E. Kir

This CHART is to shew y̆ Degrees of the
Variation of y̆ Compass as they were Observ'd
in y̆ Year 1700. in y̆ Atlantick and Indian
Ocean; and you will see in y̆ Map y̆ Varia-
tions of y̆ Compass markt over y̆ Great South
Sea, as they were found in 17 1/8.

NORTH

AMERICA

Louisiana

G. of Mexico

N. SPAIN

T. FIRMA

GREAT

S. AME-
RICA.
Brazile

SOUTH

SEA

Magellans stra
Ter. del Fuego

New Found
Land

WEST VARIATION

Bermudas

The Line of no Variation

Cuba
Jamaica Hispaniola
Caribbe

Equinoctial L.

Peru

C. S. Augustin

Pacifick sea

Chili

LaPla-
ta

EAST VARIATION

Degrees of Variation

Ireland
G. Britain
France
Spain

Azores I.

Barbary

Canary I.

Tropick of Cancer

EUROPE

AFRICA

Guinea

C. Verde

I. Ferdinando

S. Matheo

Ascension

S. Helena

I. Trinidada

Tristan di
Cunha

C. Lopas

C. Good
hope

I. Denia

WEST VARIATION

First Meridian from London

S. Paulo

Degrees of Variation

Persia

Arabia

Maldiva

C. Garasfui

C. Ambra

Madagascar I.

C. S. Sebastian

ASIA

INDIA

Ceylon

Siam

CHINA

N. Holland

PARTS

UNKNOWN

Straits of
Annian

C. Blanco

S.te Se-
bastian
New Albion

C. Men
docino

CALIF

Mozeemleck

Many Villages

Gna

Many Mo
on y̆ Islands

GULF of

Parts
High
Mountains

Unkn

t is difficult for historians not to become involved with maps. At a minimum, maps show where historical events unfolded and how those places were perceived in past eras. But as a historian of cartography, I have learned that maps can reveal so much more. For example, they also may be able to help explain the more important question of why events in history happened as they did, where they did. As the quotation on page vi by the Renaissance English mathematician Thomas Blundeville (c. 1522–c. 1606) suggests, maps are fascinating and complicated records that take some getting used to and at least some preparation on the part of viewers to be fully appreciated. I have worked with maps for more than forty years, and I continue to be amazed by their complexity.

Blundeville's words of more than four centuries ago allude to the principal purpose of this book, to increase the reader's insights into and knowledge of maps and the history surrounding them. A map gives graphic display to spatial relations and is a subjective, sophisticated communications system that functions not only in its own time but also across time with historical evidence from the past. It is composed of numerous formal elements, including the main image of the featured geographic area, the cartouche, and the scale, and such seemingly less formal elements as secondary illustrations, ornamentations, and commentaries—all designed to work together to get the cartographic message across to the viewer. A map operates on many different levels of perception—geographical, historical, scientific, and technological—as well as functioning artistically and aesthetically. Thus a map can be viewed and considered from a variety of perspectives.

In the pages that follow, I hope to foster a greater grasp and enjoyment of maps by looking at the often amazing and beautiful diversity of the less formal map elements on examples of European and American cartography dating from the late fifteenth century to the late nineteenth century. These elements can be admired in and for themselves as well as for how they function on their maps. Especially on printed maps, they also served as invaluable, trusted graphic sources of information about new worlds for a largely illiterate Old World viewing public.

I am grateful to many individuals and institutions in this country and abroad, too numerous to list here, for their generous help during my research. I am above all indebted to coordinator Ann Hodges and her staff, especially cartographic archivist Ben Huseman, of the Department of Special Collections and the Virginia Garrett Cartographic History Library of the University Libraries at the University of Texas at Arlington; to Ralph Ehrenberg, chief of the Geography and Map Division of the Library of Congress; to the former chief of the Geography and Map Division of the Library of Congress, John Hébert, and his staff; and at the Newberry Library in Chicago to James Akerman, director of the Hermon Dunlap Smith Center for the History of Cartography, and Robert Karrow, retired curator of special collections and curator of maps, and his staff. At Sterling Publishing, I am grateful to my editor, Barbara Berger; production manager Sal Destro; cover designer Jason Chow; and prepress administrator Elana Mitchel. I would also like to thank Joseph Gonzalez and Perri DeFino at gonzalez defino, ny, and their associate Susan Welt, for packaging the book. Any and all shortcomings this volume may have are, of course, of my making and not theirs.

—Dennis Reinhartz
Santa Fe, New Mexico

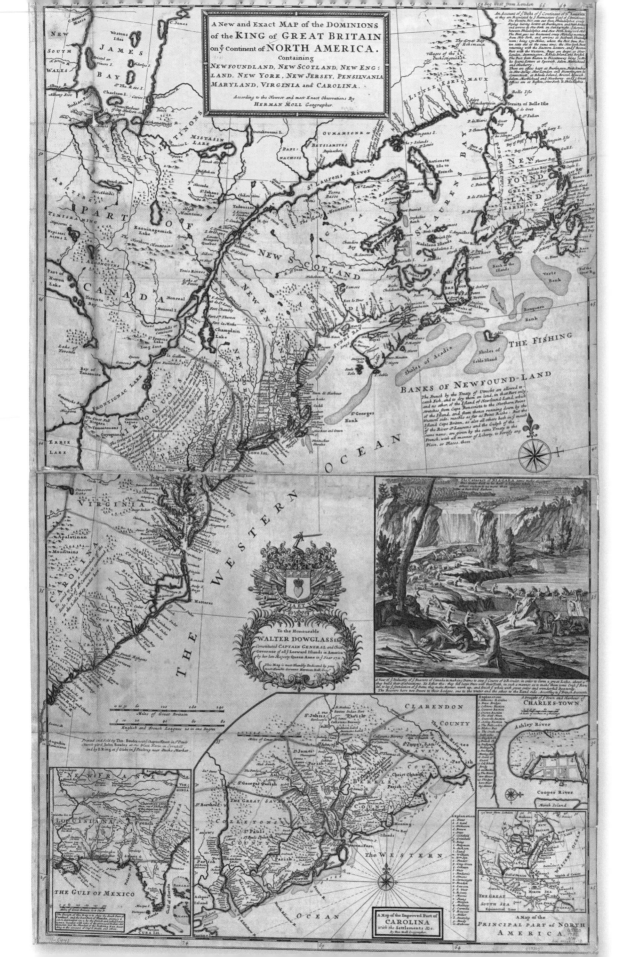

Herman Moll, A New and Exact Map of the Dominions of the King of Great Britain . . . (Beaver Map), *1715.*

What Is a Map?

"As maps become less strange to us they grow more wonderful. So we take them into our homes to make ourselves more at home in the universe."

—David Greenhood, Mapping, 1964[1]

ost people know what maps are, how they work, and how to use them. Or do they? Maps have been around a long time, and they have become a ubiquitous and necessary part of our everyday lives. From road maps to weather maps, maps shown on the news, maps used in advertising, and even online virtual street-view maps, maps have become so common that they are mostly taken for granted. But that was not the case in earlier times.

A map graphically displays spatial relations on the earth and elsewhere across the known universe and in the realms of fiction and fantasy. Even the simplest maps are complex communications devices that transmit messages—information that can be perceived one way by contemporary map users and another way by later viewers.

All maps are multidisciplinary, mathematically based pictorial records that are regularly populated by symbols for precisely designating mountains, fields, towns, buildings, fortifications, and more. They are two-dimensional representations of edited three-dimensional realities. They exist at a place where art and science meet. As such, maps are collectively an important part of human history and an equally important body of sources for understanding it.

Maps are as old as language. They epitomize "graphicacy," the ability to communicate with pictures, which dates back further than "literacy," the ability to read and write. Beyond geographical information, maps communicate data that is scientific and technological, cultural and social, economic and political, as well as historical. The term "map" derives from the material on which the map's images were recorded: the Latin word for "napkin" was *mappa*, which the ancient Romans appropriated from Phoenician Carthage, where the same word meant "signal cloth," referring to the semaphore flags used to communicate at sea. "Cartography" is the English translation of the nineteenth-century French word *cartographie*—in French, *carte* means "map" (derived from the Latin *charta*, a leaf or sheet of paper), and *graphie* is a system of writing or study (derived from a Latinized Greek word for "to write").

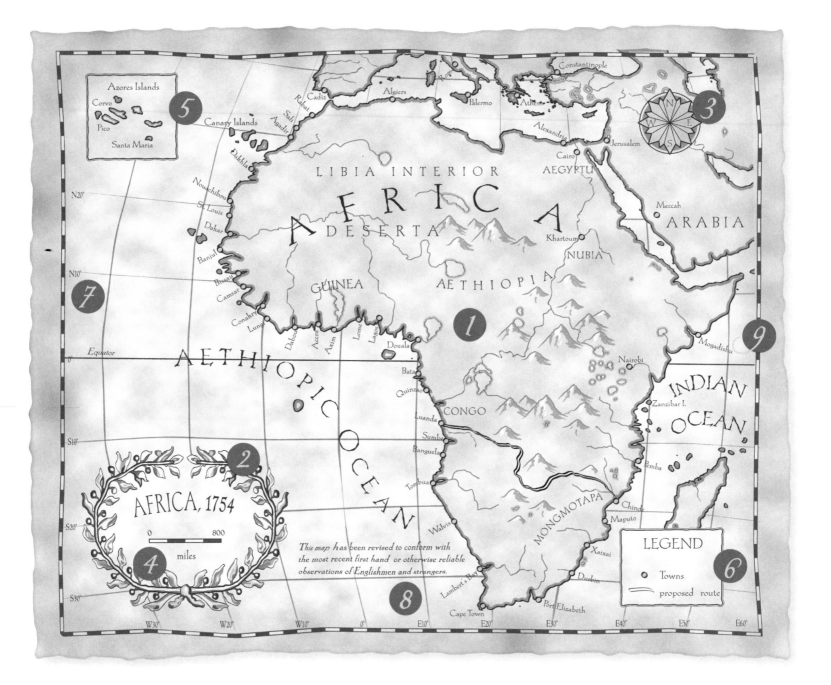

1 Subject Matter

2 Cartouche

3 Compass Rose

4 Scale

5 Inset

6 Legend

7 Coordinates

8 Commentary

9 Neatline

The practice of mapping is an ancient one, developed throughout history by different peoples for different reasons. Examples painted on or carved into stone or scratched on mammoth bone, recording local environments and dating back more than ten thousand years, have been recognized by scholars, while probably other artifacts, even older and more alien to us, go unidentified because we are not able to decipher them. From early times and at least into the nineteenth century, people have admired—and sometimes even stood in awe of—the knowledge maps convey and the efforts of those who made them and used them.

Grasping what a map has to say, the messages it has to convey, requires deconstructing it to understand how it works. At the outset, we must ask some basic questions: What is the subject matter of the map? For whom was the map made? How was it produced? When? Why? How was it to be used? Achieving some understanding of a map's context also is helpful: Where does the map fit in the past? What or who were its sources of information? And for what other maps did it serve as a source or inspiration? The answer to each question provides pertinent information about the map and its contents. Once the map has been thus comprehended, its workings are easier to figure out.

The anatomy of a map is complicated. Maps are composed of a number of different parts that on a good exemplar work together to deliver its messages effectively, as illustrated in the diagram on the opposite page. Below and on pages xviii–xxi enlarged details from the Beaver Map are used to illustrate basic map elements.

BEAVER MAP, SUBJECT MATTER CARTOUCHE. First, there is the **subject matter** or area depicted—that is, what the map is all about. On the famous so-called Beaver Map by the British cartographer Herman Moll (1654?–1732), published in London and Dublin in 1715–c. 1754 (see full map on page xiv), for example, the title, *A New and Exact Map of the Dominions of the King of Great Britain on ye Continent of North America . . .* , along with the main land-mass image, describes the subject matter as the British North American colonies in the early eighteenth century.

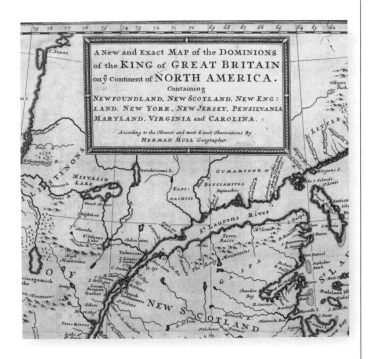

BEAVER MAP, DEDICATION CARTOUCHE. Cartouches, as demonstrated on the Moll map, are labels, sometimes elaborate, that may contain a map's title, author, date and place of publication, sources, and/or dedication to a sponsor. Moll's title cartouche (*top right*) is located in the center at the top of the map. Below the middle of the map is a second cartouche dedicated to William Dowglass Esqr., a would-be Whig governor (he was never appointed) of the Leeward Islands in the Caribbean Sea; it is an ornate cartouche topped with Dowglass's family crest. Although it's not the case with this map, cartouches are often decorated with people, plants, animals, and assorted other design devices. As we will see, these design devices are often used in other areas of a map as well.

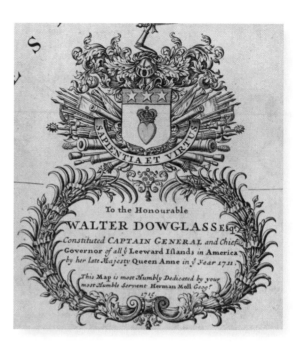

BEAVER MAP, COMPASS ROSE. Compass roses or **wind roses** (cartographic compass faces that show the cardinal directions), **arrows**, and **symbols** such as the fleur-de-lis (sometimes used to indicate north) help the map user become oriented to the directions of the map. More modern maps, especially those created since the eighteenth century, typically are oriented with north at the top. Earlier maps were often oriented toward holy cities. For example, medieval Christian world maps have Jerusalem at the center and east at the top; Islamic maps indicated the direction of Mecca, which Muslims face during prayer. Some were formatted with the west or east of a specific landmass at the top of the map, based on sailors' approaches to those lands from the sea. And many portolan navigational sea charts (see page 5) were meant to be viewed from all four sides on a map table. A relatively simple but effective compass rose (*right*) can be seen at the center, near the right side, of the Beaver Map.

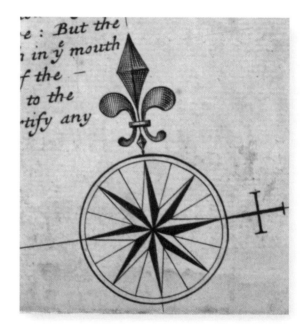

BEAVER MAP, SCALES. Similarly, maps generally have **scales** that allow the user to compute distances on them. Moll's map has two of them (*below*), for "Miles of Great Britain" and for "English and French Leagues 20 in One Degree," in the lower left corner off the coast of "Carolina."

The Cataract of NIAGARA, some make this Water Fall to be half a League while others reckon it no more than a hundred Fathom.

A View of ỹ Industry of ỹ Beavers of Canada in making Dams to stop ỹ Course of a Rivulet, in order to form a great Lake, about wch they build their Habitations. To Effect this: they fell large Trees with their Teeth, in such a manner, as to make them come Cross ỹ Rivulet, to lay ỹ foundation of ỹ Dam; they make Mortar, work up, and finish ỹ whole with great order and wonderfull Dexterity. The Beavers have two Doors to their Lodges, one to the Water and the other to the Land side. According to ỹ French Accounts.

BEAVER MAP, INSET. Some maps include **insets** that are inserted enlargements of important parts of the subject matter or that show the subject matter as part of a yet larger area. Moll's map has four such insets at the bottom, showing the southeastern part of North America, "Carolina," "Charles-town," and the British colonies as part of North America. To the right of the Dowglass cartouche is the famous inset of beavers building a dam near Niagara Falls, shown here, the source of the map's nickname.

BEAVER MAP, LEGEND. A map may also include **legends** that explain the symbols used for physical features, towns, roads, and other contents of a map. On the lower right of Moll's map a detailed legend delineates the parts of the fortress in the "Charles-town" inset, and another legend shows the European settlements in the "Improved Part of Carolina." The Beaver Map is so well drawn and efficiently labeled that it does not require an overall legend.

Explanation
1. Johnsons Raveline
2. Draw Bridges
3. Colletons Bastion
4. Carteret Bastion
5. Craven Bastion
6. The Half Moon
7. Granville Bastion
8. Ashley Bastion
9. The Pallisados
10. Blakes Bas
11. The Creek on both sides
12. English Ch.
13. French Ch.
14. Presbyterian Meeting H.
15. Ana-baptist Meeting Hou.
16. Court of Guard
17. Coll Rhetts Bridg
18. anotherRed
19. The Ministers House
20. The Quakers Meeting House

A Draught of y̆ Town and Harbour of
CHARLES-TOWN.

10 20 30 40   80   120
A Scale of 120 Paces or 3 Furlongs

Ashley River

Cooper River

Marsh Island

BEAVER MAP, COORDINATES. Maps can contain systems of **coordinates** of longitude and latitude, frequently marked along their borders and/or superimposed over them as grid systems. A grid of coordinates allows the mapmaker to locate places accurately and helps the viewer see the subject area's location relative to the equator and east or west of a mathematically determined baseline or prime meridian running through Greenwich (London), Paris, Rome, Washington, D.C., or elsewhere. Moll was British; the prime meridian of the Beaver Map is Greenwich, which he indicated with the phrase "Deg West of London," in the right upper border set along the longitudinal degrees.

60   59 Deg. West from London   55   5

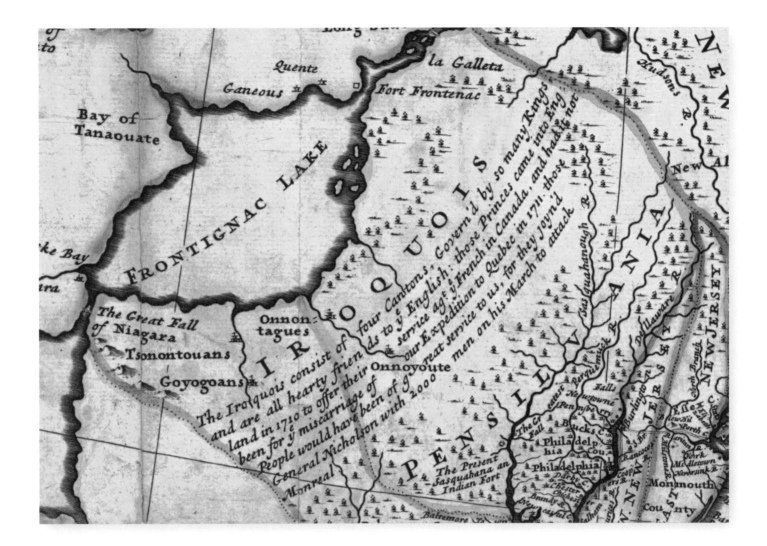

BEAVER MAP, COMMENTARY. Moll's cartography is renowned for its **commentaries**. These are written descriptions on the face or within the borders of a map that discuss aspects of, for example, the geography, history, and politics of its subject matter. Several of the Beaver Map's commentaries are particularly interesting, including one located in western "Pennsylvania" at the left center of the map about the "Iroquois" Indians.

BEAVER MAP, NEATLINE (LOWER LEFT CORNER). Upon completing their maps, cartographers usually framed them with an outermost border known as a **neatline**. Whether fancy or simple, the neatline encloses the imagery and all other parts of a map; it is where the subject matter and messaging of a map terminate.

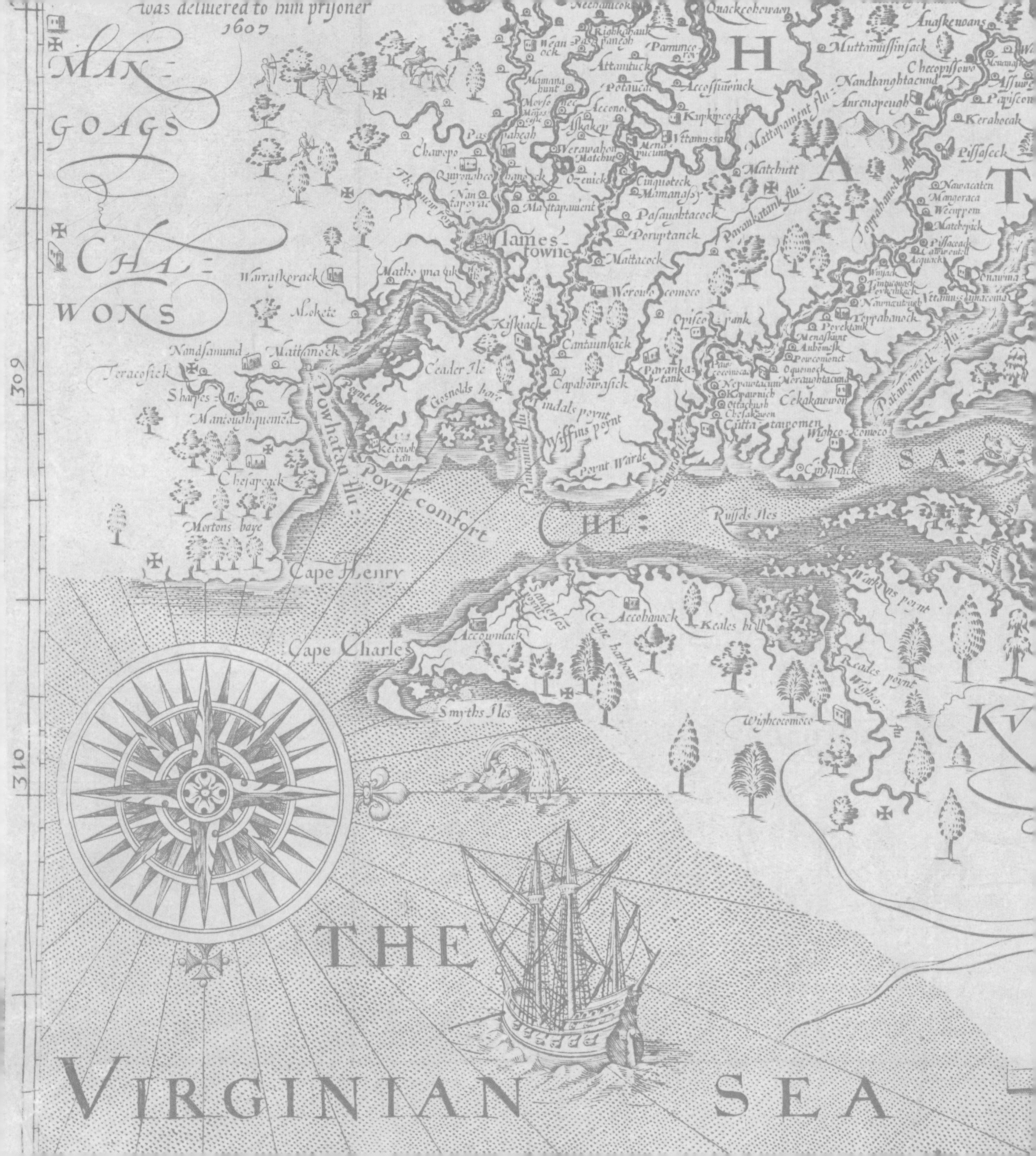

was deliuered to him prijoner
1607

MAR=
GOAGS

CHL=
WONS

H

A

T

Nechanicok
Quackcohowaon
Anaskenoans

Righkahauk
Pamunce
Muttamussinsack
Mewawaf
Checopiffowo
Weanock
Attamtuck
Nandtanghtacund
Assure
Papifco
Mamana hunt
Potaucac
Accoffiwinck
Anrenopeugh
Kerahocak
Morfos
Mesos
Accono
Askaky
Kupkipcock
Pissaseck
Chawopo
Werawahon
Matchut
Mena pucunt
Vttamussak
Pasipahegh
Nawacaten
Mongoraca
Weoppem
Matchepick
Quiyonghcohanock
Ozenick
Cinquoteck
Mamanassy
Matchutt
The Quein Cask
Nantaporac
Pasaughtacock
Pissaseck
Laurent
Acquack
Mattapanient
Poruptanck
Pamunkey flu:
Jamestowne
Matchacock
Pavankatank flu:
Onawma
Matho yna ok
Mattacock
Terrapkanock
Werowocomoco
Winjack
Timpquough
Tewkokhack
Neimautauin
Vttamussunacomoa
Waraskoyack
Mokete
Kifkiack
Opifco pank
Pevekhank
Terrahanock
Menaskant
Anhemesk
Peurcomet
Oquoinock
Nandsamund
Mattanock
Cantaunkack
Nepawtacun
Korawnich
Merawghtacum
Cekakawwon
Teracosick
Chesapeck
Ceader Ile
Capahowasick
Pavanka tank
Ottachnah
Chelakuen
Cutta tawomen
Wighco comoco
Sharpes Ile
Gosnolds hope
Tindals poynt
Nyffins poynt
Cinquack
Manconahquemea
Powhatan flu:
Poynt hope
Poynt comfort
Poynt Warde
S E A
Mortons baye
Kecoughtan
Russels Iles
Warkns poynt
Cape Henry
CHE
Reades poynt
Wighco
Sanderes poynt
Accowmack
Accohanock
Keales hill
Cape harbour
KV
Cape Charles
Smyths Iles
Wighcocomoco

THE

VIRGINIAN SEA

Compass Roses, Wind Faces, Commentaries, and Other Additions

*"History . . .
is exceedingly difficult to follow without maps . . . and, it may be whispered,
geography untouched by the human element is dull to an extraordinary degree,
duller even than mapless history, and that, the Dodo said,
was the driest thing that it knew."
--Colonel Sir Charles Close, president of the British Royal Geographical Society,
1927-30, The Map of England, 1932 [2]*

As demonstrated in the Introduction, no matter how skillful a cartographer was in facilitating the communication process, a map's content, context, and infrastructure are often complicated, and no detail of a map should be dismissed as merely decorative or otherwise superfluous without careful investigation. Some details may prove to be unimportant, but typically each element that appears on a map can act as a significant component of its delivery system. Compass roses, wind faces, insets and commentaries, coats of arms, crests, flags, and dedications are all elements that can provide crucial information to map users.

Compass Roses

For most maps to function properly, they must include an element known as a compass (or wind) rose, which displays the cardinal directions and orients a viewer to the map's spatial relationships. Because the four directions had traditionally been named for the winds by the ancient Greeks—Boreas (north), Notus (south), Eurus (east), Zephyrus (west)—the earliest compass roses, unfolding like a flower, were referred to as *wind roses*.[3]

The compass rose not only helps the map user determine where he or she is or how to journey from place to place, but also indicates where one site is located in relationship to others. The term *orientation*, when it us used for determining one's location in reality or on a map, derives from the fact that frequently early maps had east (in Latin, *oriens*) at the top instead of north, which is how most maps are laid out today. Situating north at the top of maps did not become established as the predominant cartographic convention until the nineteenth century.

Mosaic compass rose at the Monument to the Discoveries in Lisbon.

Over time, compass roses—unlike scales, which were never really ornamental but always essential—became artistic as well as scientific additions to maps; a synthesis of function and ornament, they grew to be quite embellished and decorative. Some of the finest and most beautiful examples of compass roses were painted on portolan charts. By the late seventeenth century, compass roses had faded from most maps, but today they decorate everything from architectural features, such as the enormous mosaic rose in the square beneath the Monument to the Discoveries in Lisbon (*above*), to items of clothing. Perhaps the ultimate testimony to the decorative appeal of compass roses is cartocravatia (neckties decorated with cartographic imagery); one eye-catching example is a silk tie (c. 1980) from the Mystic Seaport Museum in Connecticut, covered with colorful compass roses lifted from the charts in its extensive collection.

PREVIOUS PAGES: *Right*, **TABULA ANEMOGRAPHICA SEU PYXIS NAUTICA.** Jan Jansson, Amsterdam, c. 1650. *Left*, **VIRGINIA— DISCOVERED AND DISCRIBED BY CAPTAYN JOHN SMITH.** Oxford, 1612.

The Compass

The appearance of compass roses on maps coincides with the evolution of compasses. The earliest compasses—magnetized pieces of metal or ore suspended in water or by silk thread and pointing in a constant direction—first appeared in China, perhaps as early as the fourth century BCE. They likely were not brought farther west, via overseas and caravan trade routes, until almost 1,500 years later, in the twelfth century CE. [4] In the tenth century, Vikings are believed to have used rudimentary, sundial-like compasses in the Baltic and North seas and the north Atlantic Ocean[5], but eleventh- and twelfth-century Italian voyagers were the first Europeans we know of who used magnetized devices in navigation, in the Mediterranean Sea. At the end of the thirteenth century, the first forerunner of the modern compass was produced in Renaissance Italy. It consisted of a magnetic needle pivoting in a box over a painted diagram showing the four cardinal directions and their geometrical subdivisions.

This illustration of a mariner's compass is from Epistola de Magnete, *1269, an early treatise on magnetism by thirteenth-century French scientist Pierre de Maricourt.*

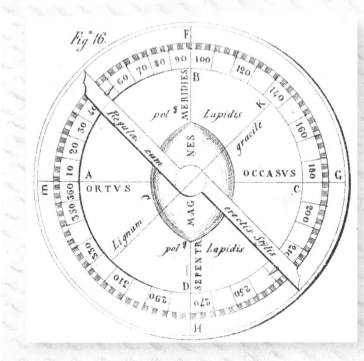

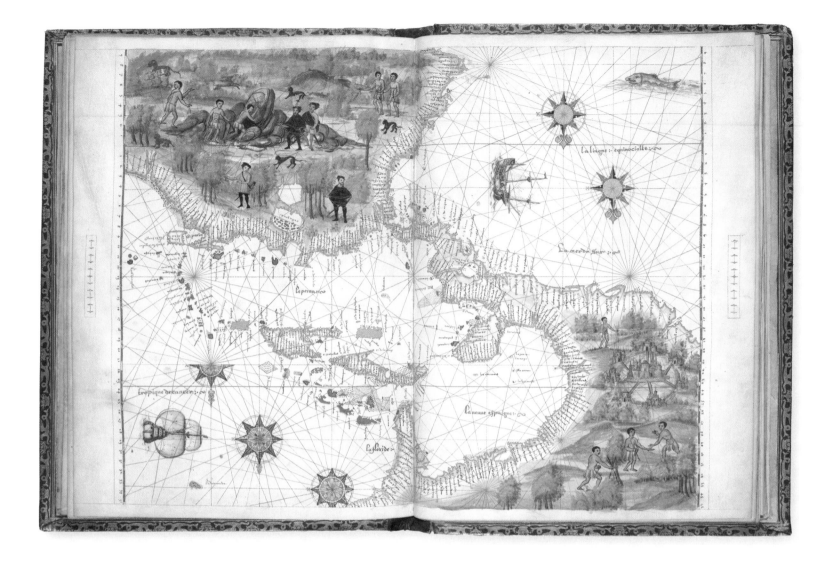

VALLARD ATLAS, CHART 10, WEST INDIES, MEXICO, CENTRAL AMERICA, NORTHERN SOUTH AMERICA. The *Vallard Atlas* (1547), created by a group of cartographers from Dieppe, France, of whom little is known, contains many wonderful examples of compass roses on its various hand-drawn charts (see page 58). The south-oriented (south at the top) chart here, centering on the Caribbean Sea and the Gulf of Mexico, has four and a half of the roses—all done in shades of red, blue, and gold—indicating north with one of two versions of a fleur-de-lis. The two and a half main roses in the Atlantic are more ornate, with actual flowers at their centers, and a bit larger, probably because of considerations of style and emphasis. Clearly, the half rose was drawn that way so that it would not encroach upon the portrayal of the islands at the juncture of the Greater and Lesser Antilles. The two compass roses in the Pacific are somewhat plainer but equally functional.

Portolan Charts and Other Navigational Maps

The origin of compass roses on maps is intimately connected with the development of medieval navigational maps known as portolan charts, where the roses functioned as nexuses of rhumb lines, or directional navigation lines (more on these below). Using a map's compass rose, a navigator could align his map with an actual compass placed over the map.

The use of portolan charts dates back to the late thirteenth century; they were likely developed in the Mediterranean region with the advent of the compass. (Note that the word "map" usually refers to cartography depicting land areas, while "chart" refers to cartography depicting sea, air, or space.) It is believed that since classical times, navigators kept pilot books, mostly for coastal areas, known as *portolani* that recorded notes on everything, including hazards to avoid, harbor features to look out for, sailing directions, distances between ports, and the cycles of tides. Originally, portolan charts were graphic extensions of the *portolani*.

The earliest portolans accurately portrayed portions of the shorelines of the Mediterranean and Black Sea basins. Navigators constantly updated their portolan charts with new information from journeys undertaken as well as from other shared or captured charts; the various sources were combined and recombined to provide the most accurate renditions for users on future voyages. The tradition of portolan-style sea charts lasted well into the seventeenth century.[6]

Anonymous portolan chart of the Mediterranean Sea, c. 1320–50, probably drawn in Genoa.

To facilitate a navigator's understanding of the location of various sites and the relationships between them, portolan charts used rhumb lines (a rhumb line follows a single compass bearing, and crosses all meridians of longitude at an equal angle); these navigational lines emanated from multiple compass roses (or linear "starbursts" representing roses) typically placed in the blank spaces of the open seas. Anywhere from eight to thirty-two points of direction radiate from the roses to create a network of intersections, angles, and geometric quadrants that extends across the map. Rhumb lines, compass roses, and other directional devices, including fleurs-de-lis, not only helped sailors plot courses; they also assisted cartographers in drawing and making truer copies of portolan charts.

VIRGINIA—DISCOVERED AND DISCRIBED BY CAPTAYN JOHN SMITH. It did not take long for compass roses to migrate from sea charts to other types of maps. On land maps, mapmakers quickly shortened the rhumb lines and eventually dropped them as they were less necessary for orientation. A good example of a transitional compass rose can be seen on Captain John Smith's map *Virginia*, published in Oxford in 1612. Smith (1580–1631) was an English explorer of Virginia and the Chesapeake Bay from 1607 to 1609 and of New England in 1614. His three-month survey of the area surrounding the Jamestown settlement in Virginia was the major source for the map, which was actually drawn by John Tappe in 1608, engraved by William Hole, and published by Joseph Barnes. The map is oriented to the west (west is at the top), as indicated by the perfectly functional thirty-two-point compass rose (*above*) placed in the blank space of the Atlantic in the lower left corner. Rhumb lines extend from the compass, but only as far as the land so as not to interfere with the more noteworthy depiction of topography and place.

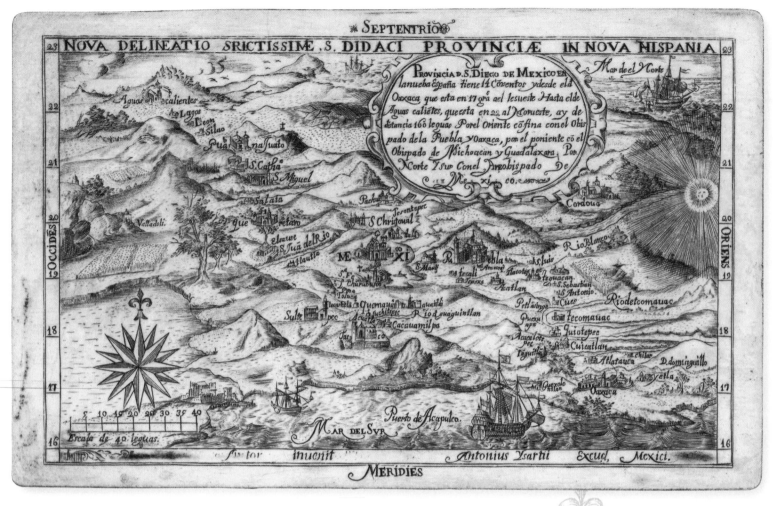

PROVINCIA D[E] S. DIEGO DE MEXICO EN LA NUEBA ESPAÑA . . .

On this c. 1682 map by Mexican engraver Antonio Ysarti, done in a
somewhat less decorative style than the Smith map, the sixteen-point
compass rose placed in the bay at the lower left does not even have
rhumb lines. During the rationalist Age of Enlightenment in eighteenth-
century Europe, the content and functionality of maps became more
important than style and traditional artistry, and compass roses were
wholly simplified.

A MAP OF THE WEST-INDIES OR THE ISLANDS OF AMERICA IN THE NORTH SEA . . . On Herman Moll's map of the West Indies from *The World Described . . .*, his seminal atlas published in London from 1709 to 1754, the compass at the right center of the map (shown enlarged at left) near the neatline is small and basic. It has only eight points, orients the map to the north, and is essentially nondecorative. Eventually, in the nineteenth century, with the general acceptance of the convention of orienting a map to the north, compass roses ceased to be a part of many maps.

Wind Faces

Another element of early maps, especially those of the sixteenth century, was the wind face. Related to compass roses in function and ornamentation, wind faces usually were portrayed as cherub heads, sometimes surrounded by clouds, blowing wind out of their mouths. The faces and the winds they produced were positioned and labeled in order to help navigators and other map users for whom knowledge of wind direction was important. As compasses came into greater use, wind faces began to pass from cartography. While they were included on maps far into the seventeenth century, in this later period the faces served increasingly as decorative elements.

PORTOLAN ATLAS OF NINE CHARTS AND A WORLD MAP . . . Some of the most beautiful and serviceable wind faces circle the earth on the cartography of Battista Agnese (1514–64), a Genoese chart maker who worked in Venice. On this alluring example from 1544, twelve wind faces surround an unexpectedly correct map of the world that indicates the track of Ferdinand Magellan's first global circumnavigation of 1519–22. The winds represented here by faces are all named after classical Greek and Roman wind divinities. For example, in the west is "Favonius v[e]l Zephir," after the Roman god Favonius and his Greek counterpart, Zephyrus, bearers of favorable light breezes to fill the sails of ships sailing across the Pacific Ocean. In the northeast is "Aqvilo vel Boreas" for the Roman Aquilo and the Greek Boreas, who bring cold winter air to places such as northern Russia. And in the south is "Avester vel Notvs" for the Roman Auster and the Greek Notus, who send late-summer storms.

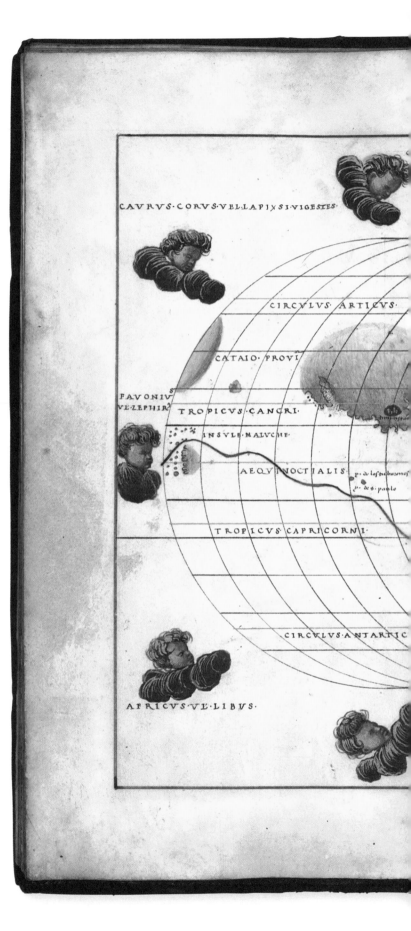

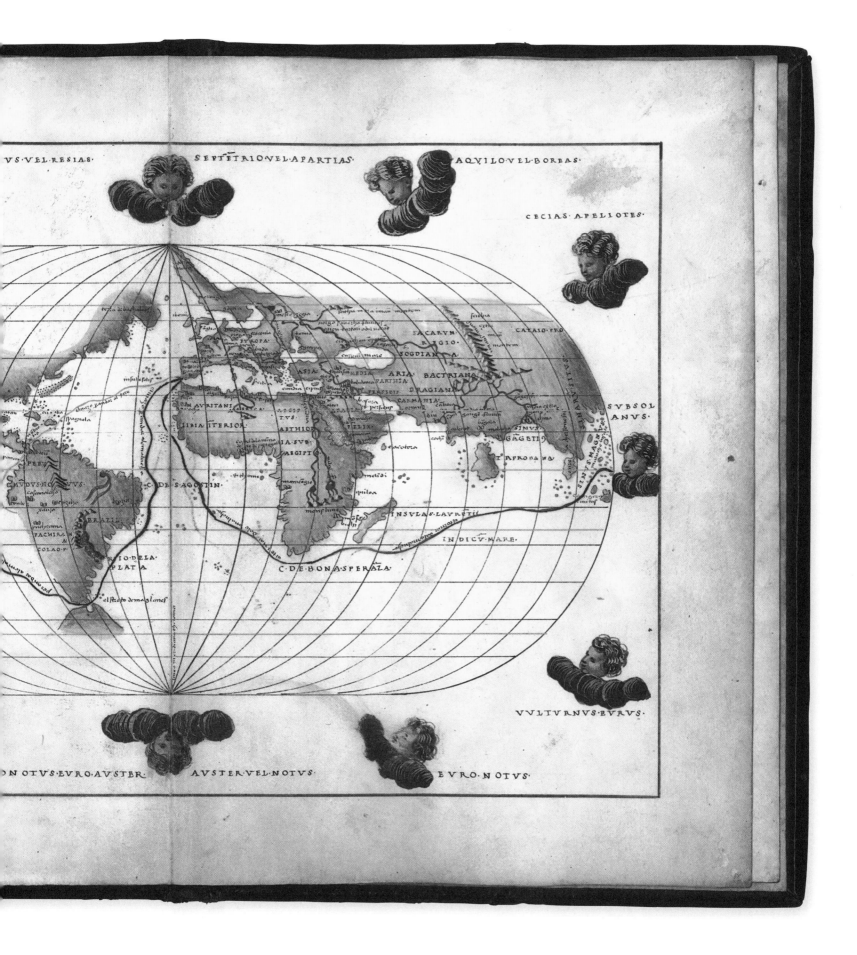

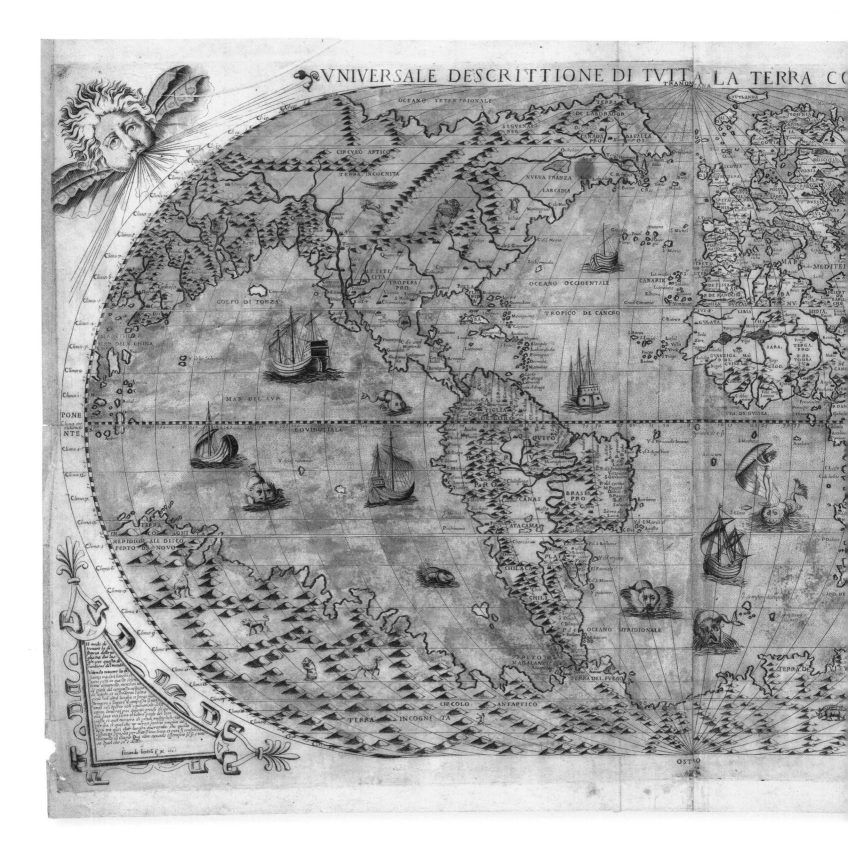

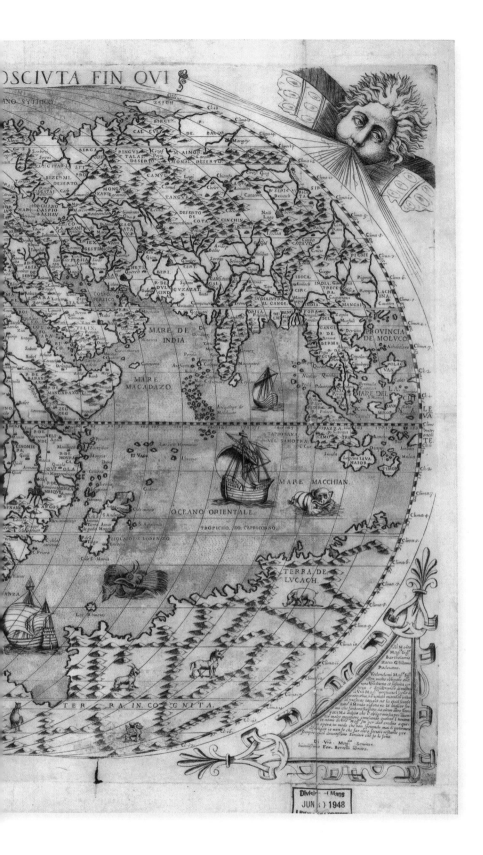

VNIVERSALE DESCRITTIONE DI TVTTA LA TERRA CONOSCIVTA FIN QVI. On this detailed world map by Paolo Forlani (active 1560–76), which was engraved on copper and published in Venice in 1565, there are only two unnamed wind faces. They gently peer over the upper left and right corners of the map and are testimony to the existence of winds and weather around the world rather than representing specific zephyrs. (The wind face from the west is seen enlarged above.) Neither the Agnese nor the Forlani maps have compass roses; most entire world maps were oriented to the north.

Insets and Commentaries

Insets and commentaries both enhance maps and help them deliver their messages. Insets usually are smaller details that have been expanded from the larger subject matter or are pictures pertaining to it; they are included for emphasis, explanation, or aesthetics. Commentaries function similarly but less graphically; they are composed of text.

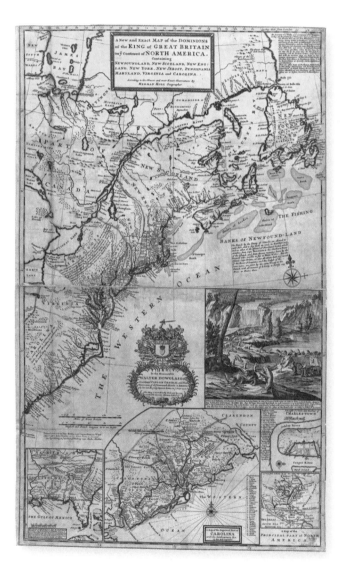

HERMAN MOLL. *A New and Exact Map of the Dominions of the King of Great Britain on ye Continent of North America . . .* (Beaver Map), 1715 (see larger version of full map on page xiv).

BEAVER MAP, INSETS. On Moll's Beaver Map (see full map, *left*), in the lower left corner an inset map (*opposite, top left*) of southeastern North America appears to the south and west of the British colonies. In the first half of the eighteenth century, this borderland between the Carolinas and Florida was an area of contention between not only the British and the Spanish, but the French as well. In the lower right corner there is an inset map of Charleston (see page xx), a British colonial port that was growing in economic importance in the trade in sugar and molasses, slaves, and rice. Directly below it is an inset showing the European divisions of North America (*opposite, top right*).

BEAVER MAP, COMMENTARY. The largest inset, at the bottom center of the map, is of South Carolina; Moll put the inset there as part of a design decision. Under the word "Carolina" in the main map, he comments (*opposite, bottom left*), "To avoid to [sic] great a Contraction of the Scale, Part of Sth. Carolina is continued in the Little Map under here." Design problems aside, the prominence given by Moll to South Carolina in the insets is directed to contemporary European viewers of the map, British and otherwise, and intended to strengthen the case for the British claims to the area.

BEAVER MAP, COMMENTARY. In another instance of the role this map played as an implement of British imperialism, to the north in the Atlantic, under the label "The Fishing Banks of Newfound-Land" (*opposite, bottom right*), Moll includes a more lengthy commentary asserting British rights to these valuable cod-fishing grounds over those of the French.

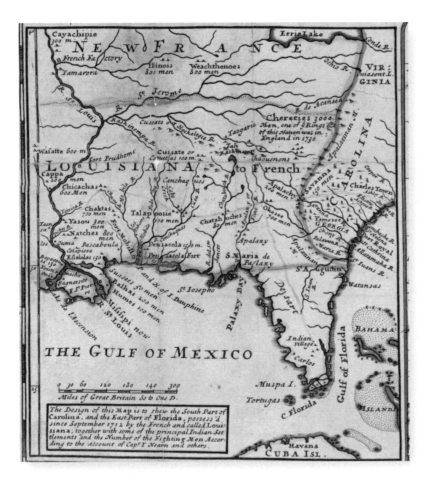

THE GULF OF MEXICO

0 30 60 120 130 140 300
Miles of Great Britain 50 to One D.

The Design of this Map is to shew the South Part of Carolina, and the East Part of Florida, possess'd since September 1712 by the French and called Louisiana; together with some of the principal Indian Settlements and the Number of the Fighting Men According to the account of Capt. Nearn and others.

A Map of the

PRINCIPAL PART of NORTH

AMERICA. 1731.

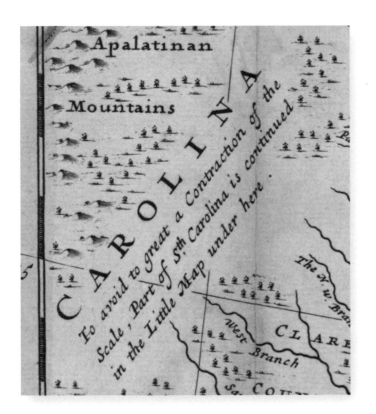

Apalatinan

Mountains

CAROLINA

To avoid to great a Contraction of the Scale, Part of Sth Carolina is continued in the Little Map under here.

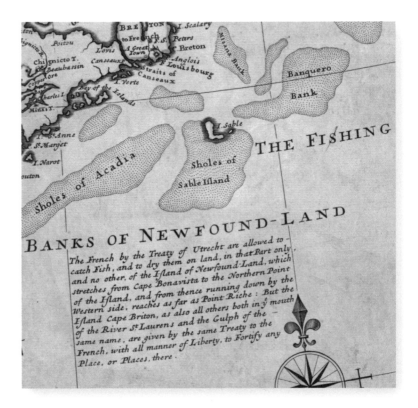

THE FISHING

BANKS OF NEWFOUND-LAND

The French by the Treaty of Utrecht are allowed to catch Fish, and to dry them on land, in that Part only, and no other, of the Island of Newfound-Land, which stretches from Cape Bonavista to the Northern Point of the Island, and from thence running down by the Western side, reaches as far as Point Riche: But the Island Cape Briton, as also all others both in y mouth of the River St. Laurens and the Gulph of the same name, are given by the same Treaty to the French, with all manner of Liberty, to Fortify any Place, or Places, there.

NIEUW A
op t Eylant

A. Het Fort B. de Kerck C. de Wintmolen D. dese Vlagge wert op gehaelt als daer Schepen in de Haven ko

303   304   305   306

NOVI BELGII NOVAEQUE ANGLIAE . . . Insets also regularly contain views of cities that are of as much interest to students of history today as they were to the map users when the maps were first issued. A good example is found on *Novi Belgii Novaeque Angliae nec non Partis Virginiae Tabula . . . ,* a map of the east coast of North America from Virginia to Canada, originally engraved by Flemish cartographer Nicolaes Visscher (1618–79) in 1655 or 1656 and republished in color in 1685 by Peter Schenk (1660–1718). As part of the decorative cartouche there is an inset of New Amsterdam (before it became English and was renamed New York in 1674) at the bottom of the map (*below*). Prominent in the foreground are the gallows, stocks, and other facilities for the punishment of all types of transgressors. Schenk was an Amsterdam map publisher and seller whose work reflects a pride in Dutch economic might and social order as reflected in international trade with such places as New Netherlands in North America.

Traditionally, there has been a symbiosis between discovery, exploration, exploitation of resources, and cartography. As demand grew for more and better maps not only to show the way and record places but also to help European venturers access desirable resources, the maps in turn stimulated further discovery, exploration, and exploitation. The various centers of the map trade, including Antwerp and Amsterdam, developed with the extension of empires. The independence of the United Provinces of the Netherlands from Spain and the Holy Roman Empire was officially recognized in the Peace of Westphalia, a series of treaties signed in 1648 that ended the Thirty Years' War. By the time Schenk's map was published, the Dutch Republic—little more than three and a half decades old—had become a major commercial power that conducted trade around the world. It is estimated that in the late seventeenth century 20,000 or more ships sailed under the Dutch flag.[7]

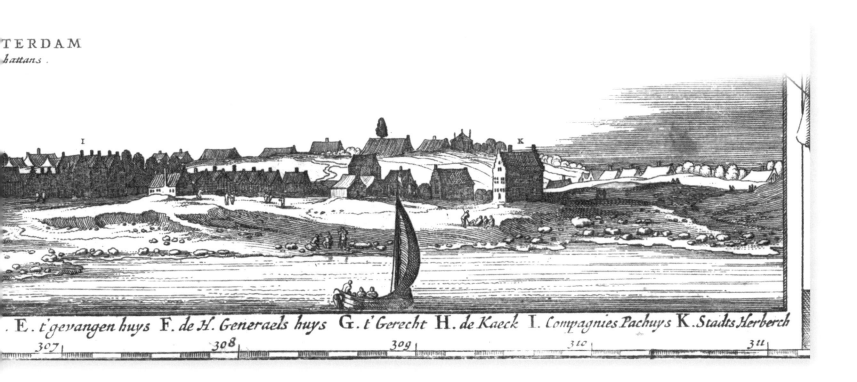

TERDAM
hattans .

. E . t'gevangen huys F. de H. Generaels huys G. t'Gerecht H. de Kaeck I. Compagnies Pachuys K. Stadts Herberch

307   308   309   310   311

CARTE DE L'AMERIQUE SEPTENTRIONALE . . . Jean Baptiste Louis Franquelin (1653–c. 1725)—the French royal hydrographer (sea-chart cartographer) and the official cartographer for New France under Louis XIV—spent many years working in Canada; he completed his substantial map of North America, *Carte de l'Amerique Septentrionale* . . . , in 1688. It served as a foundation for numerous French maps of the area published over the next quarter century. In its lower right corner is a beautiful and detailed inset of Québec (*opposite*), founded as a colonial outpost by Samuel de Champlain in 1608 on the St. Lawrence River. It even shows several French ships, anchored in the river, waiting to take on cargoes of furs and forest products for the European markets.

QUEBEC
Comme il se voit du côté de l'Est

TABLE
Alphabetique pour
connoistre les lieu
les plus remarquables,

a. La Rade ....   i. le Fort ....
b. Cul de Sac ....   t. les Ursullines ....
c. Platte-Forme ....   m. l'Eglise des Iesuittes ....
d. Place Royalle ....   n. les Classes ....
e. Effigie du Roy ....   o. la Catedralle et Parrel ....
f. Rüe a la haute Ville ...   p. Vieux Seminaire ....
g. l'Euesche ....   q. Seminaire neuf ....
h. le Simtiere ....   r. l'Hospital ....

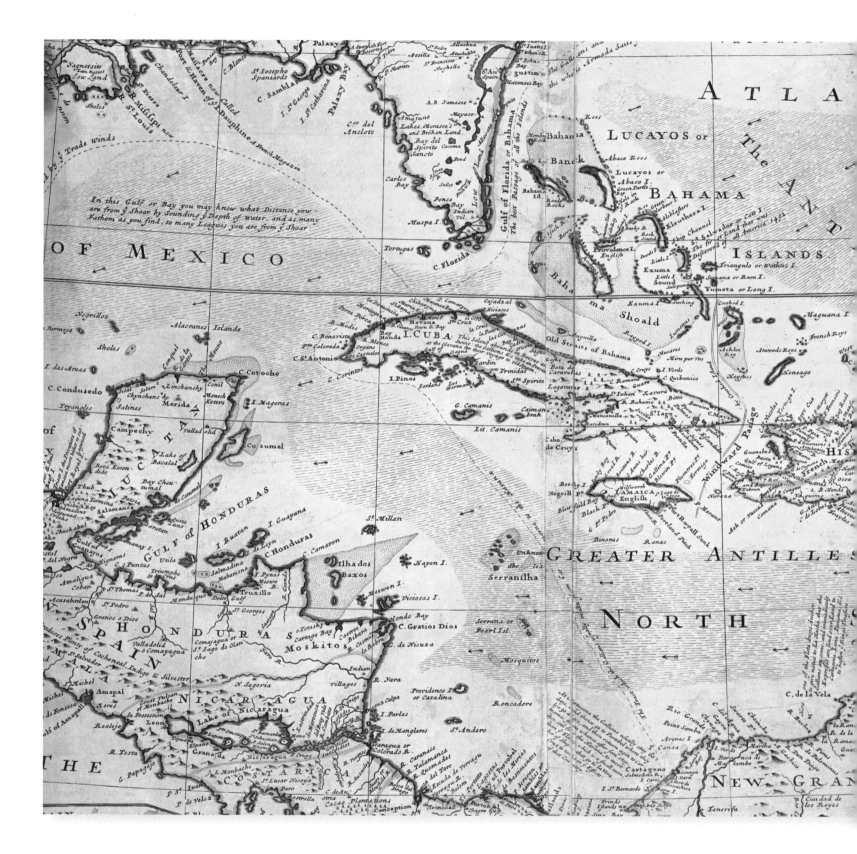

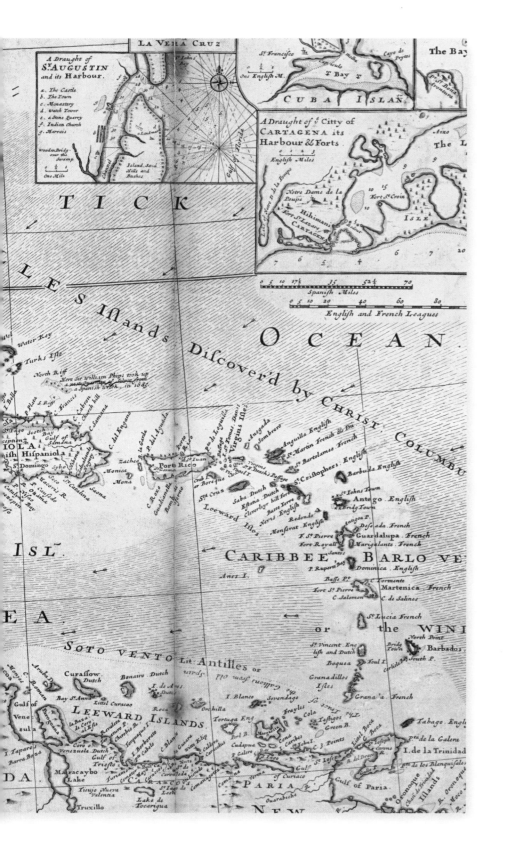

A MAP OF THE WEST-INDIES OR THE ISLANDS OF AMERICA IN THE NORTH SEA . . . , DETAIL. On this map created by Moll in c. 1715, there are "tracks" of the various Spanish treasure fleets sailing across the Caribbean from "La Vera Cruz" in Mexico to Havana, Cuba, and then on to Europe via the straits between Florida and Cuba and the Bahamas; along the northern coast of South America from the Windward Islands to Cartagena (Venezuela); and from Cartagena northward to Havana. Alongside the marked routes, Moll has placed commentaries about various ports, such as Cartagena (below).

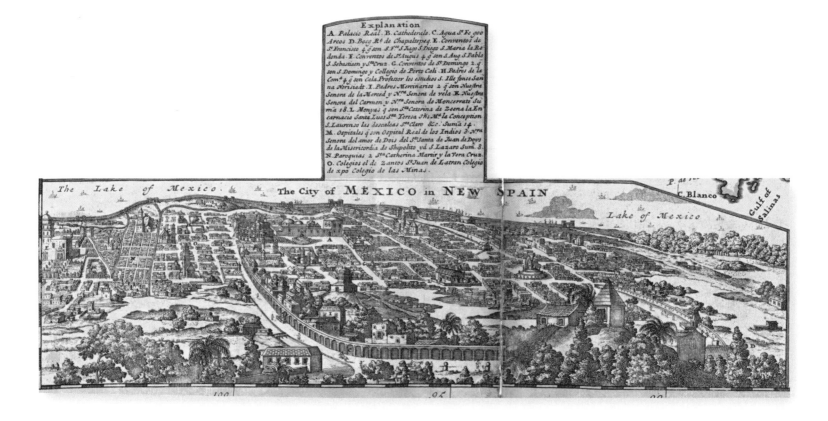

The Lake of Mexico.   The City of MEXICO in NEW SPAIN   C. Blanco   Gulf of Salinas   Lake of Mexico

A MAP OF THE WEST-INDIES OR THE ISLANDS OF AMERICA IN THE NORTH SEA . . . , INSETS. On Moll's map, there is also an inset of Mexico City, the largest and wealthiest city in the New World and the seat of the viceroyalty of New Spain (*above*). He also provides map insets of "St. Agustin" (Florida), "La Vera Cruz," "Bay & Citty of Havana" (*opposite, top*), "Bay of Portobello" (Pacific coast of Panama), and "Draught of ye Citty of Cartagena Its Harbour & Forts."

With a combination of insets and commentaries, Moll's map also underscores the Spanish treasures coming from and going to Asia via ships known as the Manila galleons; they plied the Pacific between 1565 and 1815, sailing back and forth from Acapulco to the Philippines, bringing the goods of Asia to the New World for export to Spain in exchange for the commodities of Europe and New Spain, especially the silver and gold from the mines of New Spain. Through commentaries printed above the coast of New Granada and along the western fleet course line in the Florida Strait, and with its glorious depiction of Mexico City, the map calls attention to the region's wealth and its appropriation by British privateers— private citizens with armed ships who were licensed by the government to attack foreign vessels and who prowled the Atlantic and Pacific oceans and the Caribbean Sea in search of Spanish treasure ships.

Moll was well acquainted with some of the British privateers, especially William Dampier and Woodes Rogers. Dampier circumnavigated the globe three times. Rogers captured a fabled Manila galleon off Baja California in 1710, bought himself the governorship of the Bahamas with part of the proceeds, and promptly cleared the islands of pirates, including the infamous Edward Teach, known as Blackbeard. Moll drew maps and charts for these privateers and others, and they in turn provided geographic information from the far corners of the earth—including Australia, the Solomon Islands, and East Africa—for his cartography.[8]

A. A High Tower where is always watch kept, to see if any Ships are coming from Sea, and as many Ships as many Flaggs are hung out that ye Citty may know it

Mesa de Maria Fort

Moro Fort

Water Castle

Lagide R.

A Draught of ye Bay & Citty of HAVANA.

Velha Fort

St Francisco

o molha

Cayo de Prytos

Capriculo

0   ½   1
One English M.

T Bay T

CUBA ISLAÑ.

6
6
4
4
3½
3
3

CARTE TRES CURIEUSE DE LA MER DU SUD . . .
This large, somewhat over-the-top Americas map (*below*) by Franco-Dutch cartographer Henri-Abraham Châtelain (1684–1743), printed in Amsterdam in 1719, is a veritable compendium of insets, both original and taken from other cartographic sources. Beyond the two simple functional compass roses on the equator, there are insets of cities, forts, and other geographic locations, important individuals and ordinary people engaged in daily activities, and exotic flora and fauna, as well as a number of written commentaries. Using dotted lines, Châtelain also goes further than Moll and traces routes of discovery, exploration, and trade across the Atlantic and Pacific basins.

CARTE TRES CURIEUSE DE LA MER DU SUD, CONTENANT DES REMARQUES NOUVELLES ET TRES UTILES NON SEULEMENT SUR LES PORTS ET ILES DE CETTE MER. Mais aussy sur les principaux Pays de l'Amerique tant Septentrionale que Meridionale, Avec les Noms & la Route des Voyageurs par qui la decouverte en a été faite. Le tout pour l'intelligence Des Differtations suivantes

Coats-of-Arms, Crests, Flags, and Dedications

From the fourteenth to the eighteenth centuries, many maps were liberally decked out with variations of coats of arms and with shields, crests, flags, and dedications. A coat of arms is a complete representation of the insignias and other heraldic designs, often medieval in origin, that signify a specific royal or noble family.

Heraldic insignia were much more than mere adornments on maps. Collectively, they represented the very real presence of contemporary sovereign power. They helped to demarcate and establish claims to what was being mapped. Mapmakers sometimes included insignia on maps in an attempt to lure patronage from those in power. An attractively rendered coat of arms not only beautified a map, it might also show a cartographer's support and admiration of that aristocratic family, thereby contributing to real and future sales. The coats of arms of the various European royal houses appear on many early maps.

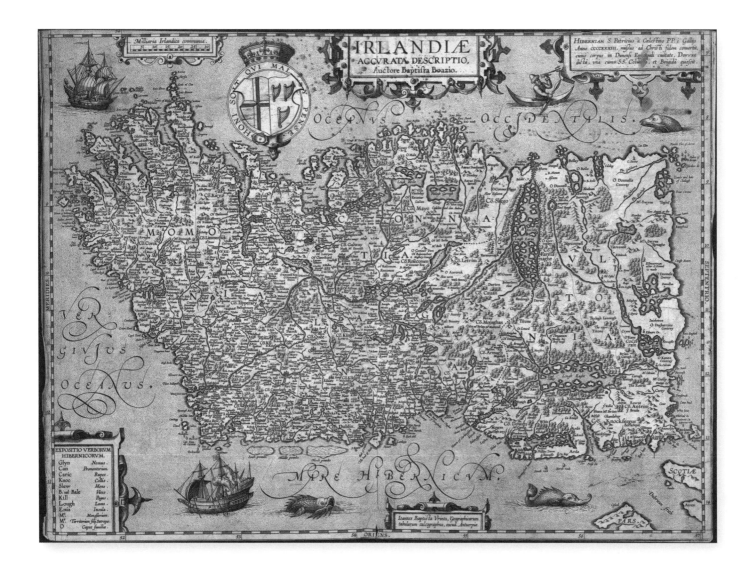

IRLANDIAE ACCVRATA DESCRIPTIO. Giovanni Battista Boazio, an Italian cartographer who worked in England from 1585 to 1606, put the Tudor coat of arms on several of his hand-colored engravings of maps illustrating places visited and claimed by Sir Francis Drake on his voyage to the West Indies in 1585 and 1586. The maps were published in a book on the voyages published in London in 1588 and 1589. Boazio's *Irlandiae Accvrata Descriptio* (c. 1606) (*opposite*), which appeared in Abraham Ortelius's *Theatrum Orbis Terrarum,* carries a variation of the coat of arms of the new Stuart dynasty (*above left*). John Smith's 1612 map of Virginia shows the coat of arms of Stuart England as well (*above right*; see full map on pages 6–7).

OVERLEAF: NOUA ORBIS TERRARUM DELINEATIO . . . Philipp Eckebrecht (1594–1667) was an astronomer and mathematician from Nuremberg, Germany, and a friend of Johannes Kepler. Eckebrecht drew an impressive map of the world in 1630, called *Noua Orbis Terrarum Delineatio . . . ,* which was published in Nuremberg in 1658. On the map, Eckebrecht's world is clutched by a massive Hapsburg double-headed eagle of the Holy Roman Empire.

LEOPOLDO ROMANORUM IMPERATO RI S.A devotissime D.D.D.

SEPTEN:

AMERICA

Circ Arcticus

SEPTEN TRIONALIS

Tropi: Cancri

MAR DEL

HORÆ SUBTRA

LINEA ÆQVATORIS

ZUR

USUS HUJUS CHARTÆ NAUTICÆ, Si ad certam horam innotuit locus lunæ ex observatione applicationis marginum ejus ad aliqua stellam notam, aut ad extremitates Solis vel umbræ terræ in locus, per remotionem parallaxium illius horæ (ubi opus est) ex viso in verum fuerit conversus: computetur ex Tabb. Rudolphi, qua hora Uramburgica locum illum verum occupaverit luna. Tunc ex pauciores horæ et minuta fuerint in observatione, differentia quæsita inter HORAS SUBTRAHENDAS chartæ, sin plures, inter ADDENDAS, deteget Meridianum loci, sub quover fatur observator: seu respondeat pictura littorû, sive discrepet: nam ea hoc medio corrigi tandem Poterit.

Vicissim, si computata est ex Tabb Rudolphi horâ Uramburgica Eclipfes lunæ, vel cujus, cunq loci lunæ veri: ea reducitur ad loca reliqua, subtrahendo vel addendo tofidem horas, quot inveniuntur ad cujusq loci Meridianum annotatæ: sed hinc fida supponitur collocatio locorum.

Cum Privilegio Cæfareo ad Annos XXX.

MERIDIES

NOVA FRANCIA

MARE ATLAN

MAR DEL NORT HENDÆ

OCEANUS

Tropicus

ÆTHIOPICUS

AMERICA MERIDIONALIS

BRASILIA

CHILI

TE AUS INC

Circulus

GVIANA

LINEA AEQVATORIS

SINGULARI RATIONE ACCOMMODATA
PHI ASTRONOMICARUM

ME PRES SAM TENEBRIS REVOCAT LEOPOLDUS IN AURAS

SEPTEN

Circul Arcticus

AMERICA
SEPTENT.
PARS
Nevado

OCEANUS CHINENSIS

Tropicus Cancri

Philippinæ Insulæ

Ilhas de Ladrones

DE NDÆ

NOVA GUINEA

MARE ARABICUM

et INDICUM

HORÆ

TARTARIA

KALKARES

SINDIOS

INDIA EXTR

MAR DI INDIA

MAR DI INDIA

Capricorni

INDIA

RA Antarcticus

Ex præscripto Tabb. RUDOLPHI pag. 33.34.35.36. et PRÆCEPTORUM LX.LXI. pag. 41.42.

Petente, typumq Emblematis innuente

IOANNE KEPPLERO MATHE MATICO CÆSAREO, amico charißimo.

ita dis posuit, suáq manu exaravit, ex sculpi denió, fide integrâ curavit

PHILIPPUS ECKEBRECHT civis Norimbergensis.

Sumptus faciente Jo. Kepplero Sculpsit Norimbergæ J.P. Walch 1630.

MERIDI

CLIMA Septentrionale
Frigida
TRI.
Temperata SEPTEN.
ZONA Torrida
MERIDI. Temperata
Frigida
CLIMA Meridionale
MERIDI

NOVI BELGII NOVAEQUE ANGLIAE . . . , LION RAMPANT. At the top of the cartouche of Schenk's *Novi Belgii Novaeque Angliae . . .* (see full map on page 16) stands the royal Dutch lion rampant—the classic heraldic pose of an animal rearing its forelegs.

CARTE DE L'AMERIQUE SEPTENTRIONALE . . . , COAT OF ARMS. Jean Baptiste Louis Franquelin's influential map of North America, reproduced in full on page 18, features a large, superbly drawn Bourbon coat of arms of the Sun King, Louis XIV.

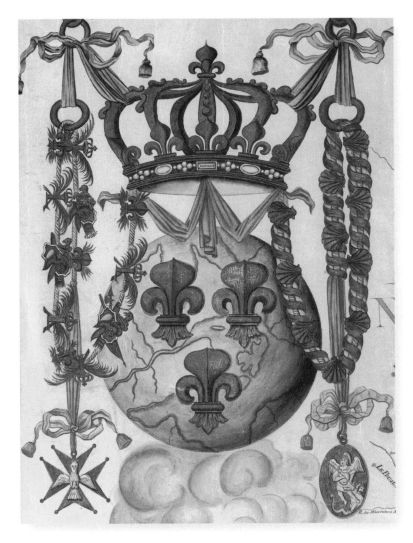

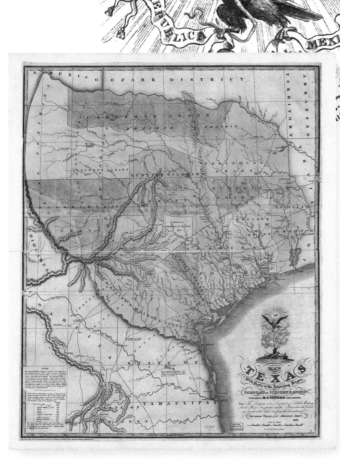

MAP OF TEXAS WITH PARTS OF THE ADJOINING STATES. A century and a half later, in the age of nationalism and revolution, the nonroyal but nevertheless sovereign coat of arms of the new Republic of Mexico appears in the cartouche (enlarged *above*) on a work by Stephen F. Austin (c. 1790–1836) entitled *Map of Texas with Parts of the Adjoining States*, published by Henry S. Tanner in Philadelphia in 1830. It consists of a prickly-pear cactus with labeled paddles (stem segments) that represent the various states of Mexico, including Texas. On top of the cactus sits the Aztec golden eagle clutching a snake in its beak. In its talons, the eagle holds a banner proclaiming an independent, republican, and sovereign Mexico, which was officially founded in 1823. Above the eagle's head is the cap of liberty (a conical cap worn by freed Roman slaves that later became a symbol of liberty), radiating freedom over Mexico.

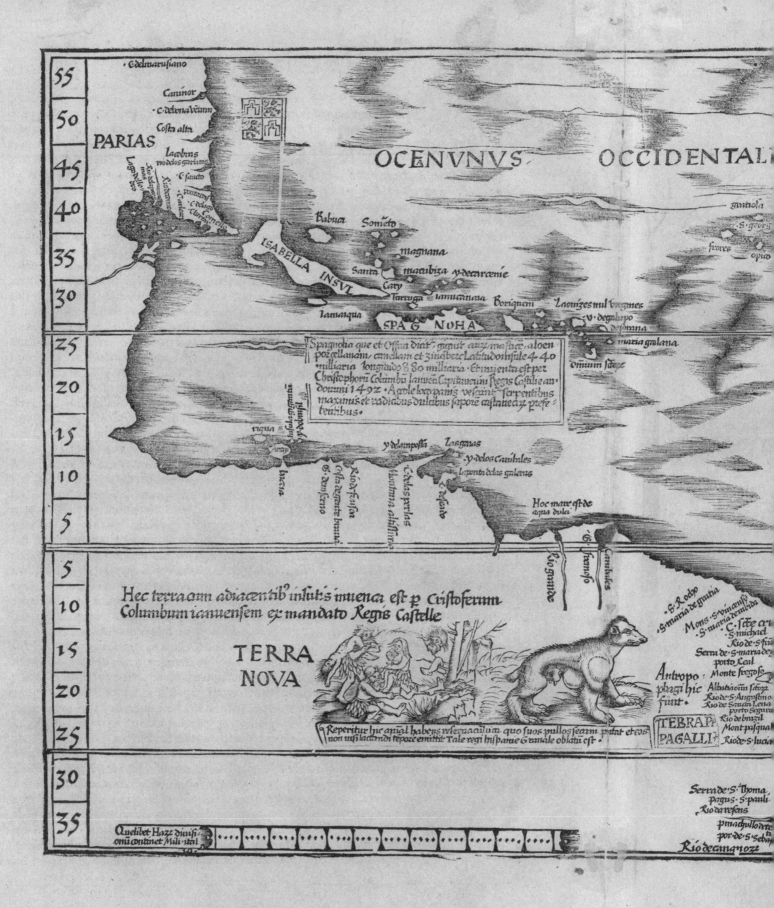

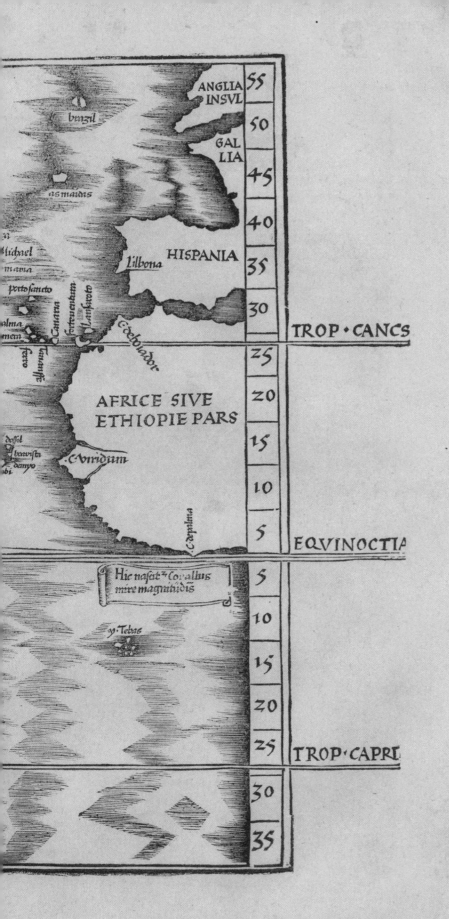

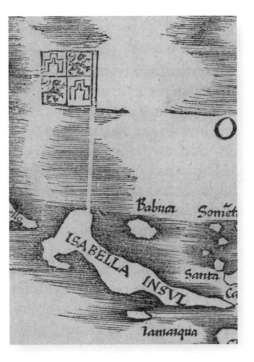

TABULA TERRA NOVA. On much earlier maps and charts, particularly those of the sixteenth century, royal coats of arms were displayed in the form of flags and banners that identified specific monarchies and the territories they claimed. On the 1523 Strasbourg edition of German cartographer Martin Waldseemüller's woodblock map *Tabula Terra Nova*, the first printed map of the Atlantic Basin—also known as the Admiral's Map—a large flag bearing the Spanish coat of arms (enlarged *above*) protrudes from "Isabella Insvl" (present-day Cuba; "insvl" stands for *insula*, the Latin word for "island").

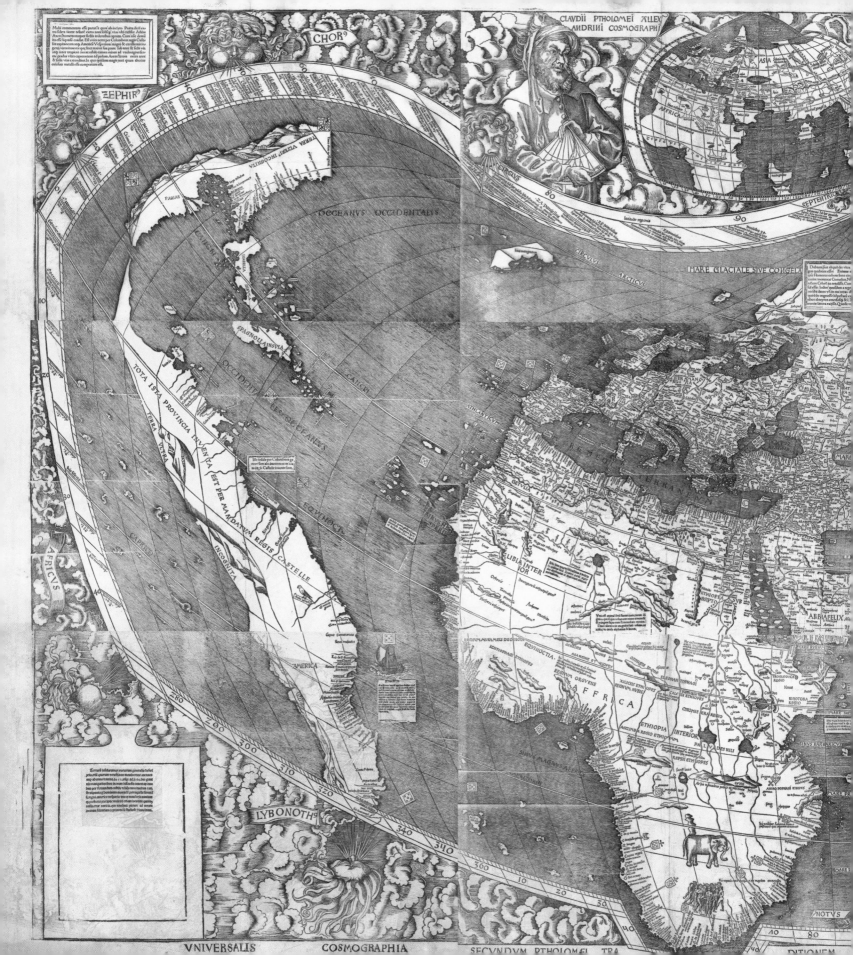

UNIVERSALIS COSMOGRAPHIA SECVNDVM PTHOLOMEI TRA DITIONEM

PREVIOUS PAGES: UNIVERSALIS COSMOGRAPHIA SECUNDUM . . . In 1507, Martin Waldseemüller (c. 1475–c. 1522) and his German humanist colleague Matthias Ringmann (1482–c. 1522) published a large, twelve-sheet map of the world, *Universalis Cosmographia Secundum . . .* , on which "America" is named as such for the first time. The 4½-by-8-foot map was the largest ever printed. The only known surviving copy of the map often called "America's birth certificate" is today on permanent display in the Library of Congress. Waldseemüller created the map, and Ringmann edited and wrote much of the accompanying text. Waldseemüller named the New World after the Italian explorer and geographer Amerigo Vespucci (1451–1512), who made at least two (although he claimed to have made four) voyages to the New World with the Spanish and Portuguese between 1497 and 1504 or so. Vespucci wrote a geographic account of his voyages in 1504, which was published in French as *Quatre Navigations* and then translated into Latin and included in Waldseemüller and Ringmann's *Universalis Cosmographia*. Apparently, the two authors were ignorant of the totality of Columbus's achievement on his four voyages between 1492 and 1504 and erroneously believed Vespucci to have been a part of the European discovery of the mainland of the New World before Columbus, and Waldseemüller, under the influence of Ringmann, designated the New World "America" on his 1507 map and others to follow, including the first edition of *Tabula Terra Nova*, in honor of Vespucci. Waldseemüller acknowledged his error on the 1522 edition and other later maps of the New World, but by that time it was too late. The name "America" was coming into more common usage by cartographers, and it stuck.[9] Amerigo Vespucci is shown at the top right of the map with an eponymous banner over his head; an enlarged detail is reproduced (*below*).

Tabula Terra Nova was first issued in 1513 as part of Waldseemüller's edition of the great classical work *Geographia* (c. 160), a handbook of geography with instructions on mapmaking, but no surviving maps, written by Claudius Ptolemy (c. 90–168). A Greek astronomer, mathematician, and geographer who was a citizen of the Roman Empire working at the great library in Alexandria, Egypt, Ptolemy is recognized as a founder of modern geography and cartography. Waldseemüller's edition of *Geographia* was published, with maps, in Strasbourg in 1513.

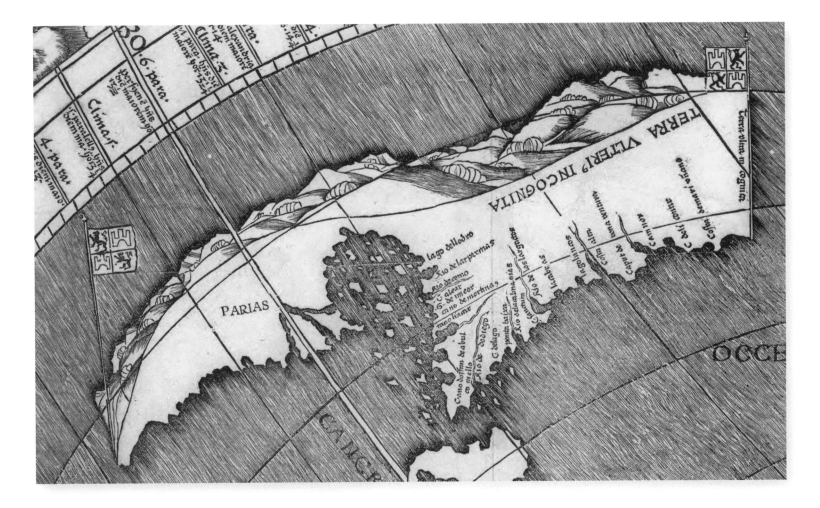

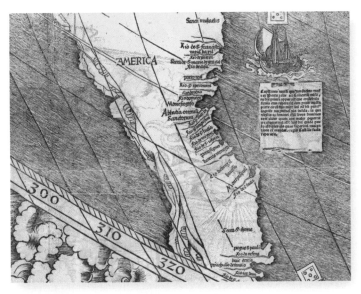

UNIVERSALIS COSMOGRAPHIA SECUNDUM . . . (NAMES AND FLAGS).
Note that Waldseemüller labeled what is now South America (*left*) as "America," probably to fill available blank space. The bottom part of the landmass of North America (*above*) is labeled *Parias*—Latin for "to make equal"—likely meaning that this part of the landmass is considered to be America as well. Parias is festooned with the Spanish flag, representing the union of Aragon and Castile (Ferdinand II and Isabella). Above "Parias" appears the phrase *Terra VIteri Incognito*, which roughly translates as "very unknown lands." Flags and banners were also used to designate the domains of enemies and others. They were the colorful devices of early European imperialism, whether real or improvised. On portolan and other sea charts, for example, Muslim coasts were indicated by banners bearing crescents and/or stars. The coasts of frequently disputed territories of rival monarchies were symbolized by similar banners bearing their coats of arms.

NOVAM HANC TERRITORII FRANCOFVRTENSIS TABULAM . . . The coats
of arms of greater and lesser nobility too were to be found on maps.
In such cases the primary purpose of the mapmakers was not to show
dominion, but probably to secure the patronage of local nobles for their
cartographic efforts. The celebrity appeal or political status conferred by
noble underwriting could also help to increase middle-class subscriptions of
maps. One of the finest maps displaying such coats of arms is *Novam Hanc
Territorii Francofvrtensis Tabulam . . .* (c. 1638), by Joan Blaeu (c. 1599–1673)
and his brother Cornelis (1610), published in Amsterdam. The subject of
the map is the major German economic center Frankfurt am Main and its
environs. The Dutch came to dominate global trade and, correspondingly,
cartography in the sixteenth and seventeenth centuries, and several
generations of Blaeus were among the leading Dutch mapmakers. In 1638, by
royal appointment, Joan succeeded his father Willem Janzsoon (1571–1638)
as the official cartographer for the Dutch East India Company. Stylistically,
in its ornateness, complexity, grandeur, and ordered presentation, this
excellently hand-colored, copper-engraved map is a prime example of the
artistic cartography of the great European Age of the Baroque.

At the very center is Frankfurt, situated on the banks of the Main River.
The city is surrounded by the forests, rivers, and towns and villages of the
Hessian countryside. On the map there are three smaller coats of arms, one
in the southwest and two in the east, perhaps locating noble estates. This
scene is then lavishly and almost completely bordered by thirty-five larger
family coats of arms of various members of the local German nobility and the
royalty of the Holy Roman Empire. These specifically identified aristocrats
underwrote the production of the map and the atlas of which it was a part,
either through subscription or outright donation. The coats of arms also
were placed around the map to underscore the importance of the city and
to induce others to purchase the map or atlas. If your family coat of arms
was not one of those shown, a blank one was intentionally left at the bottom
of the map, just to the right of the cartouche; it could be filled in with your
heraldic arms and properly designated on the individual map you purchased.

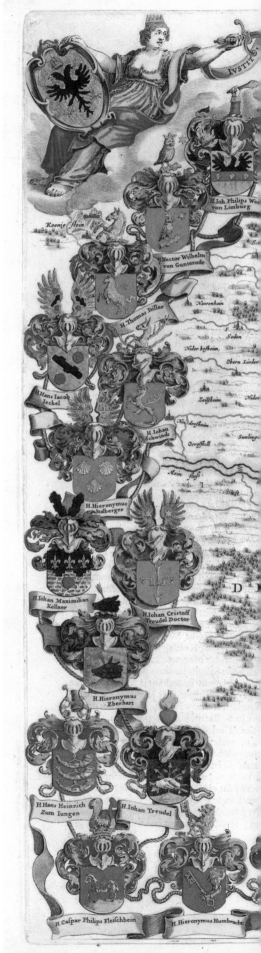

A MAP OF THE WEST-INDIES . . . , COAT OF ARMS AND CARTOUCHE. Each of Moll's large maps published in *The World Described . . .* was dedicated to an important individual and suitably marked with that person's family coat of arms. On the West Indies map (see full map on pages 20–21), there is a coat of arms of Sir William Paterson in the lower left corner. Immediately below it is a cartouche inscribed "To Wil. Paterson Esq; This Map of the West-Indies &c. Is Most Humbly Dedicated by Her. Moll Geographer." Sir William Paterson, a Scottish merchant-adventurer who founded the Bank of England in 1694, was an advocate for the unification of Scotland with England. He was also behind an unsuccessful scheme in the 1690s to seize Panama from Spain and found a Scottish trading colony in Panama.

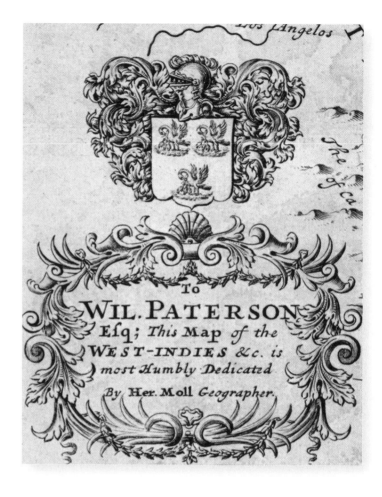

CODFISH MAP AND BEAVER MAP, COAT OF ARMS AND CARTOUCHES. Moll's c. 1717 North America map (see full map on page 60), known as the Codfish Map, is dedicated to "John Lord Sommers [Somers]," baron of Evesham and president of the Royal Privy Council under Queen Anne (see enlargement *opposite*). And his Beaver Map bears the crest of "Walter Dowglass [Douglass] Esqr." (see page xviii), who was to be appointed governor general of the Leeward Islands in the Caribbean by Queen Anne but was derailed by powerful Tories in Parliament. The fact that Moll kept the dedication on several editions of the map may point to his frugality in not wanting to re-engrave the copper plate or to engrave a new one, or it may indicate his Whig leanings. The German-born Moll was a strong advocate of the British overseas empire, as were the dedicatees on his maps of it.

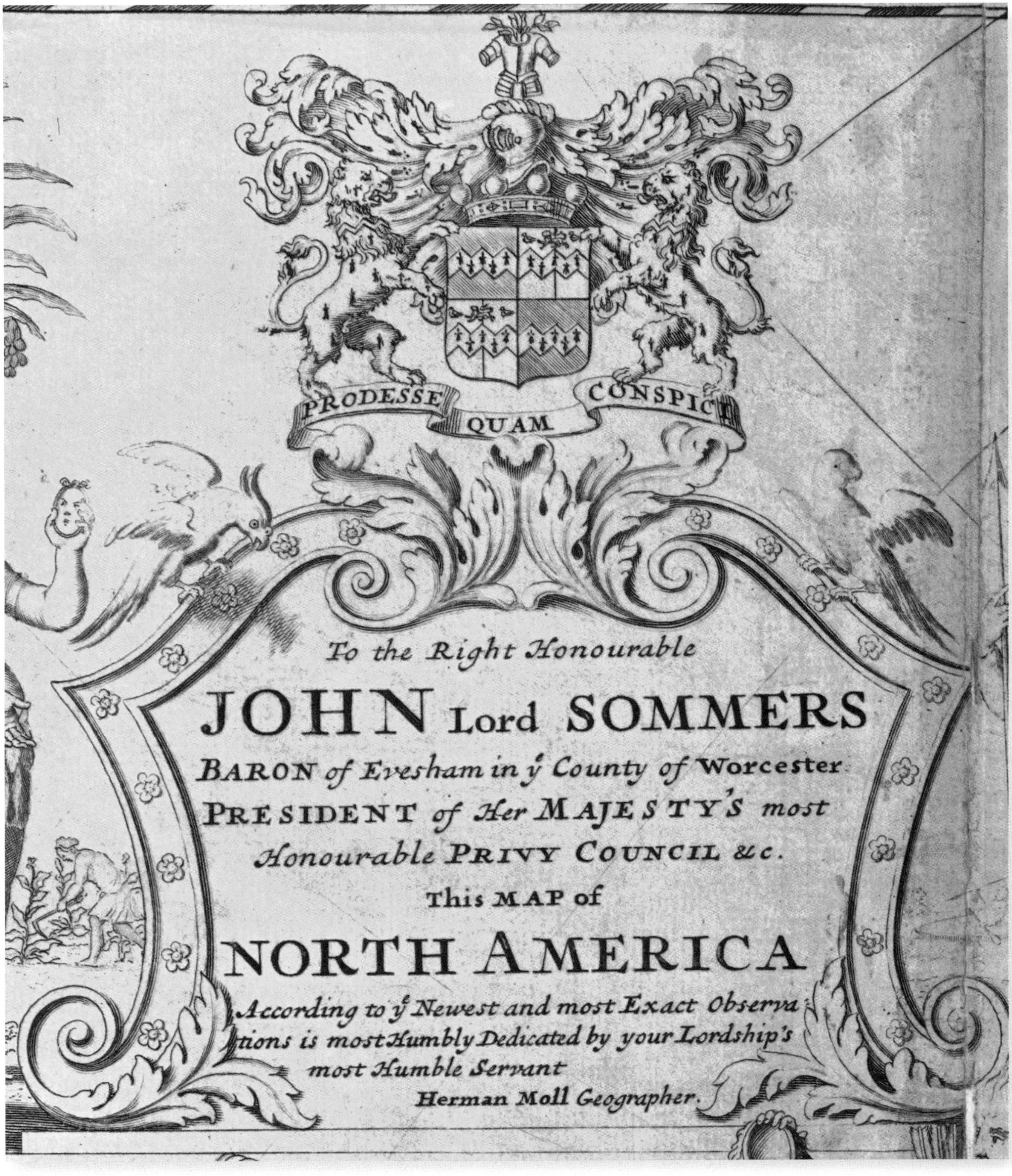

PRODESSE QUAM CONSPICI

To the Right Honourable

JOHN Lord SOMMERS

BARON of Evesham in y County of Worcester
PRESIDENT of Her MAJESTY'S most
Honourable PRIVY COUNCIL &c.

This MAP of

NORTH AMERICA

According to y Newest and most Exact Observations is most Humbly Dedicated by your Lordship's
most Humble Servant

Herman Moll Geographer.

la magnifi(que) reception du roy des moluques
faicte au sig.(r) dracke le faisant tire au port
par quater de ses galeres et luy mesme costoia(t)
les vassiau dudict drack et prenoit grand
plaisir a ouir la musique

2

All the Ships at Sea

"But thou at home, without tide or gale,
Canst in thy map securely sale,
Seeing those painted countries, and so guesse
By those fine shades, their substances;
And, from thy compasse taking small advice
Buy'st travell at the lowest price."

— Robert Herrick (1591–1674), "A Country Life" [10]

n intimate relationship between mariners, maps, and ships has existed since the early days of seafaring, and among the most common of decorative inclusions on old maps are portrayals of ships at sea. These artistically captivating images, often quite lavish, were frequently intended by mapmakers to attract attention, but their purpose often went beyond mere ornamentation.

Many of the illustrations of ships found on maps are but stylized generalizations that are meant to be purely decorative. In such instances, the images do not represent specific historic vessels but rather ships typical of the era when the map was created. The presence of such ships enhanced the visual appeal of a map and thereby contributed to its salability.

The presence of ships on maps also symbolically contributes a sense of the "high adventure" surrounding the voyages related to discovery, exploration, empire, and commerce that cartographers were called upon to illustrate. Images of these intrepid vessels helped make distant map users feel more a part of an adventure in a time before global travel; they made the perusal of maps an early form of "armchair travel." This sense of vicarious exploration continues to be a major incentive for present-day collectors of antique maps.

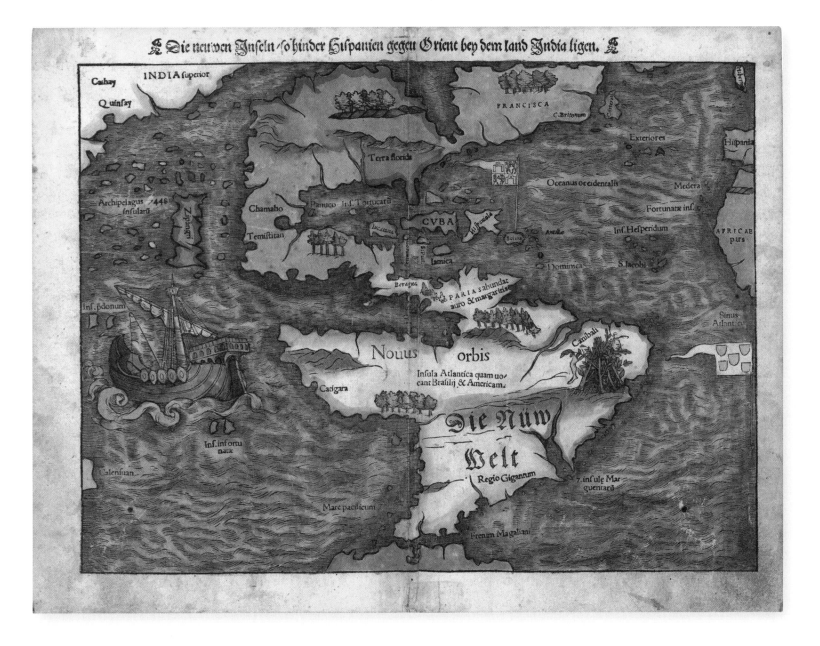

INDIA superior

Cathay

Quinsay

FRANCISCA

C.Britonum

Exteriores

Hispania

Archipelagus 448 insularū

Terra florida

Oceanus occidentalis

Medera

Fortunate insl.

Chamaho

Panuco Insl. Tortucarū

CVBA

Inf. Hesperidum

AFRICAE pars

Temistitan

Iucatana

Spagnola

Antille

Temistitan

Setina

Jamica

Dominica

S. Iacobi

Beragua

PARIA sabundat auro & margaritis

Sinus Atlanticus

Insl. pdonum

Canibali

Nouus orbis

Insula Atlantica quam uo cant Brasilij & Americam.

Canigara

Die Nüw Welt

Insl. infortu nata

Regio Gigantum

7. insule Mar guentarū

Calensuan

Mare pacificum

Frenum Magallani

PREVIOUS PAGES: *Right*, **HAVANA OP'T EYLAND CVBA.** Joan Vinckeboons, Amsterdam, c. 1639. *Left*, **LA HERDIKE ENTERPRINSE FAICT. . . .** Nicola van Sype, Antwerp?, c. 1581.

DIE NEUWEN INSELN . . . An early, classic example of a decorative nautical vessel is the stylized ship in the lower left of the woodcut New World map by German scholar and cartographer Sebastian Münster (1488–1552), from *Cosmographia*, his masterwork geographic encyclopedia of the sixteenth-century world, first published in Basel and then in numerous editions around Europe from 1540 to 1628.

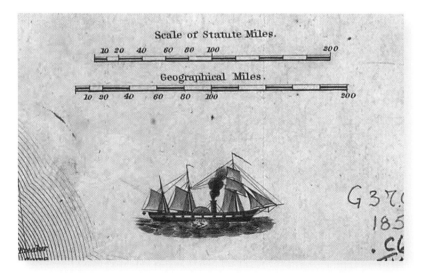

MAP OF THE UNITED STATES OF AMERICA. A much more modern ship sails eastward in a vignette below the scales on *Map of the United States of America*, issued in New York in 1850 by American cartographer and publisher J. H. Colton (1800–1893). While this is a three-master under sail, it is powered by steam as well; its smokestack billows black smoke (*left*). The image of this ship marks the period of transition from sail to steam in seafaring and epitomizes the vitality of the young, industrializing United States.

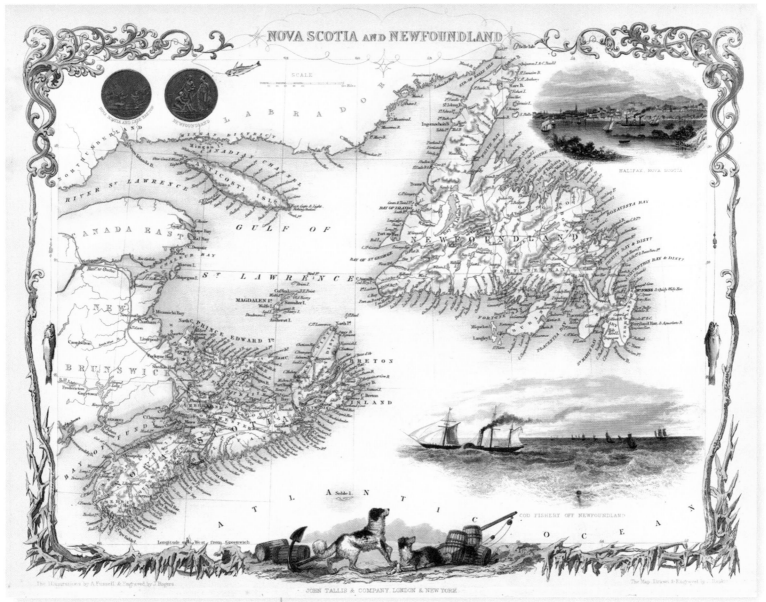

NOVA SCOTIA AND NEWFOUNDLAND

NOVA SCOTIA AND NEWFOUNDLAND. A similar sail-steamship appears in the Atlantic underway toward the mainland on the map *Nova Scotia and Newfoundland* in the *Illustrated Atlas*, published by John Tallis (1817–76) in London in 1851.

Ship Portraits

On rare occasions, the representations shown on maps were of actual vessels, such as the *Victoria*, one of a fleet of five ships with a total of 237 crewmen on board under the command of the Portuguese nobleman and navigator Ferdinand Magellan (c. 1480–1521) in the service of King Charles I of Spain (the future Holy Roman Emperor Charles V). The *Victoria* set out from Seville in August 1519 in search of a westward route to the Spice Islands—the Moluccas—in present-day Indonesia. Magellan sailed her across the South Atlantic to Brazil, then southward, and finally, in October and November 1520, he passed through the strait that today bears his name. He crossed the Pacific to the Marianas Islands, landing on Guam in March 1521, then went on to the Philippines. Magellan was killed on Mactan Island in the Philippines on April 27, 1521, during a battle led by a native tribal chieftain. Only two of the expedition's five ships, the *Trinidad* and the *Victoria*, eventually reached the Spice Islands on November 21. When the *Trinidad* began to leak, only the Victoria remained, under the command of Juan Sebastián del Cano (c. 1476–1526), to sail across the Indian Ocean, around southern Africa, and back to Spain. Only eighteen exhausted crew members were on board when the *Victoria* arrived in Spain on September 6, 1522. Cano and the survivors on the *Victoria* thus had carried out the first circumnavigation of the earth. It had taken them three years.

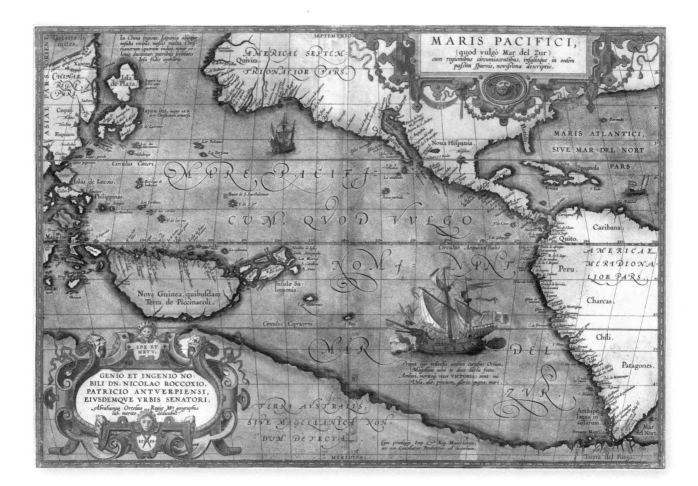

MARIS PACIFICI. In honor of the achievement of the Magellan expedition in exploring the Pacific, Abraham Ortelius (1527–98), on the magnificent first printed map of the as yet little-known Pacific Basin, titled *Maris Pacifici* (1589; *opposite*), positioned the *Victoria* prominently off the west coast of South America. Ortelius, a cartographer, cosmographer, and publisher from Antwerp, is credited with creating the first atlas, a set of maps of similar design on a shared subject matter: the *Theatrum Orbis Terrarum* (1570–c. 1644). The *Theatrum* appeared in numerous updated editions in Latin, Dutch, German, French, and English over a period of approximately seventy years, and each edition was an immediate sold-out best seller. Updated copies of *Maris Pacifici* appeared in several editions of the *Theatrum* from 1589 to 1612.

MARIS PACIFICI, *VICTORIA*. On Ortelius's map, in a presumably faithful rendition of the *Victoria* (*top right*), the ship is resplendent with billowing sails, and flags and pennants flying in the wind; her guns are firing as she begins her journey into the Pacific and Indian oceans. Members of her crew are working on the main deck; on the aft deck, someone, perhaps Cano, is taking navigational readings with an astrolabe. Up forward on the bowsprit, a protective angel apparently directs the ship on its voyage. Below the *Victoria* is an inscription in Latin identifying her and describing the Magellan expedition's passage around the world.

MARIS PACIFICI, CARAVEL. Two other ships appear on *Maris Pacifici*. Almost at the center of the upper part of the map, below a badly distended North American coastline—near where the "Rio Grande" is erroneously shown emptying into the Pacific Ocean—there is a generic caravel, a fast, small, light ship of the age of discovery (*center right*). Of unknown national origin, the caravel is in full sail, perhaps engaged in a peaceful exploration of the coast. It may even represent the *Golden Hind*, the English privateer Francis Drake's ship, which sailed along the northwest coast in the summer of 1579. Drake, too, circumvented the globe (1577–80) and was tasked by Queen Elizabeth I with finding a northwest passage between the Atlantic and the Pacific. The *Victoria* and the *Golden Hind* both became popular icons on maps of the period. Drake and other English navigators may have had copies of the *Theatrum Orbis Terrarum* with them on their voyages, and it's possible that, upon their return to Europe, they provided Ortelius with updated information for correcting his maps.

MARIS PACIFICI, GALLEY. The other ship shown on *Maris Pacifici* is in the Caribbean Sea to the northeast of the island of "La Trinidad" off the coast of South America. It looks to be a small European, possibly even Venetian-style coastal galley (*bottom right*) that could never have made the rough transatlantic journey to American waters. It is an example of a late-medieval cartographic anachronism whose function is purely decorative.

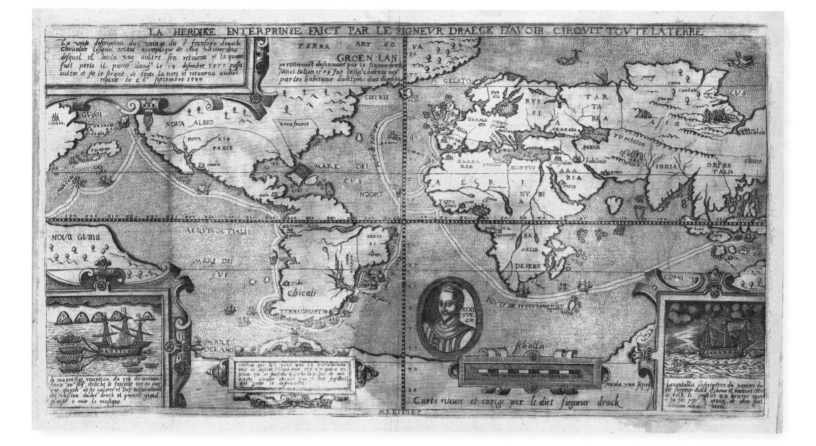

LA HERDIKE ENTERPRINSE FAICT . . . The *Golden Hind* makes another appearance in *La Herdike Enterprinse Faict par le Signeur Draeck D'Avoir Cirquit Toute la Terre* (c. 1581), by the Flemish cartographer Nicola van Sype (1589–1641). The map traces the *Golden Hind's* track around the world. In an inset picture (enlarged, *opposite*) in the lower left corner, the *Golden Hind* is being towed in becalmed waters, an actual event from the account of Drake's circumnavigation.

la magnifiq; reception du roy des moluques
fuicte au sig. dracke le faisant tire au port
par quater de ses galeres et luy mesme costoiãt
des vassiau dudict drack et prenoit grand
plaisir a ouir la musique

VERA TOTIUS EXPEDITIONIS NAUTICAE . . .

At the bottom center of another map depicting
Drake's circumnavigation of the globe, *Vera
Totius Expeditionis Nauticae: Descriptio D. Franc.
Draci . . .*" (c. 1595), by the Flemish cartographer
Jodocus Hondius the Elder (1563–1612), the
Golden Hind is portrayed at rest.

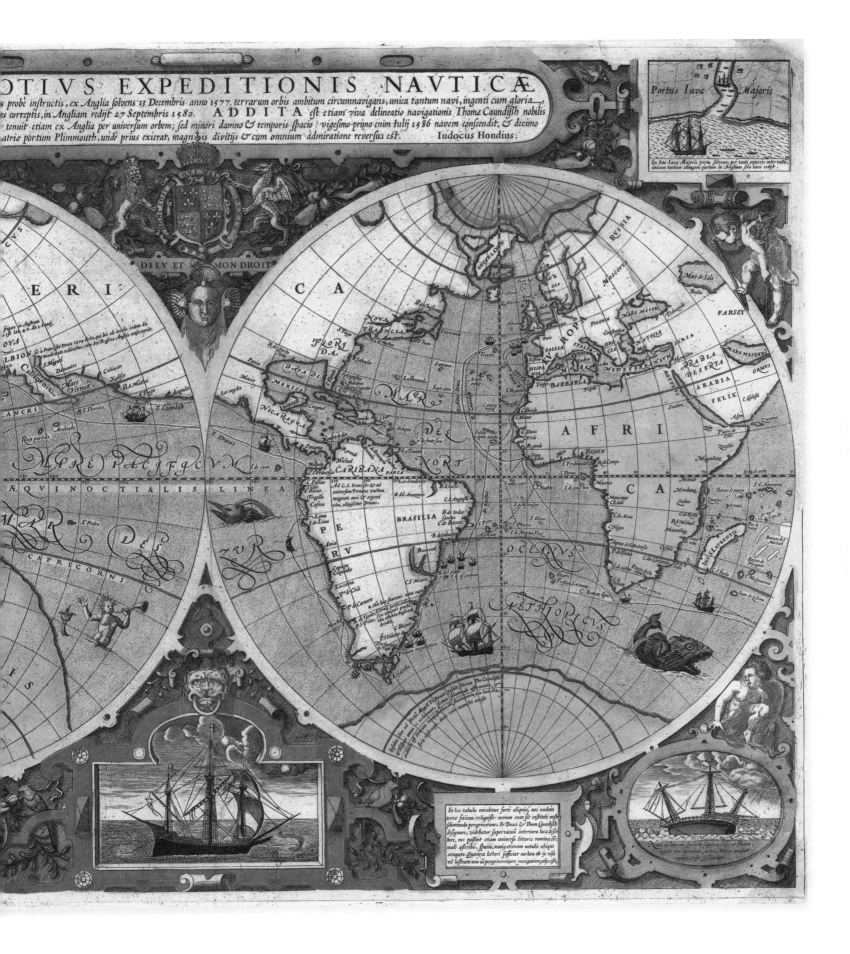

Ocean and Empire

It is difficult to make unadorned oceans look interesting on maps, especially those that are not sea charts to be used expressly for navigation; in other words, there is a lot of empty space to be filled. Thus, another design role these beautifying ship elements play on maps is to fill in (sometimes even clutter up) distracting blank spaces in vast and/or unfamiliar seas. In so doing, they help keep the attention of the map's viewer better focused on its essential content. The emphasis remains more on what is known rather than on what is unknown.

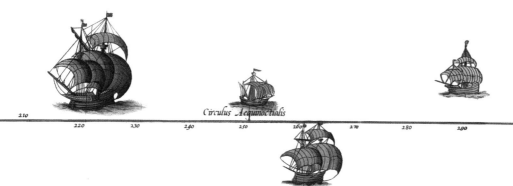

AMERICAE SIVE NOVI ORBIS . . . On the various editions of the notable Americas map *Americae Sive Novi Orbis, Nova Descriptio*, from Ortelius's *Theatrum*, the oceans are populated by generic ships of various sizes. While the ships embellish the map, contribute to its ambience, and fill its oceanic voids, they also function in other capacities. For example, some of them might represent the early fabled Manila galleons (see pages 20–22). These ships might also represent the pirates who were drawn by the promise of treasure to prey on the Manila galleons and other shipping.

The 1570 original copperplate for *Americae* was replaced in 1579, probably because it was worn. On the maps produced by the two plates, the principal differences between the three ornamental ships in the Pacific and the one in the Atlantic are minor. On the newer map, for example, the largest, westernmost ship's sails are no longer billowing, and it is now sailing eastward. But on a later edition of *Americae* (1589), the three ships in the Pacific expanded to an impressive gathering of fifteen varied period ships, while in the Atlantic one became eleven. Undoubtedly, these increases were in large part merely aesthetic, but they also may have been intended to signify the increased European exploration of the New World and its surrounding waters. The larger number of ships may therefore have been intended as a graphic political statement reinforcing the expansion of European global power and dominance for an as yet largely illiterate map-viewing audience.[11]

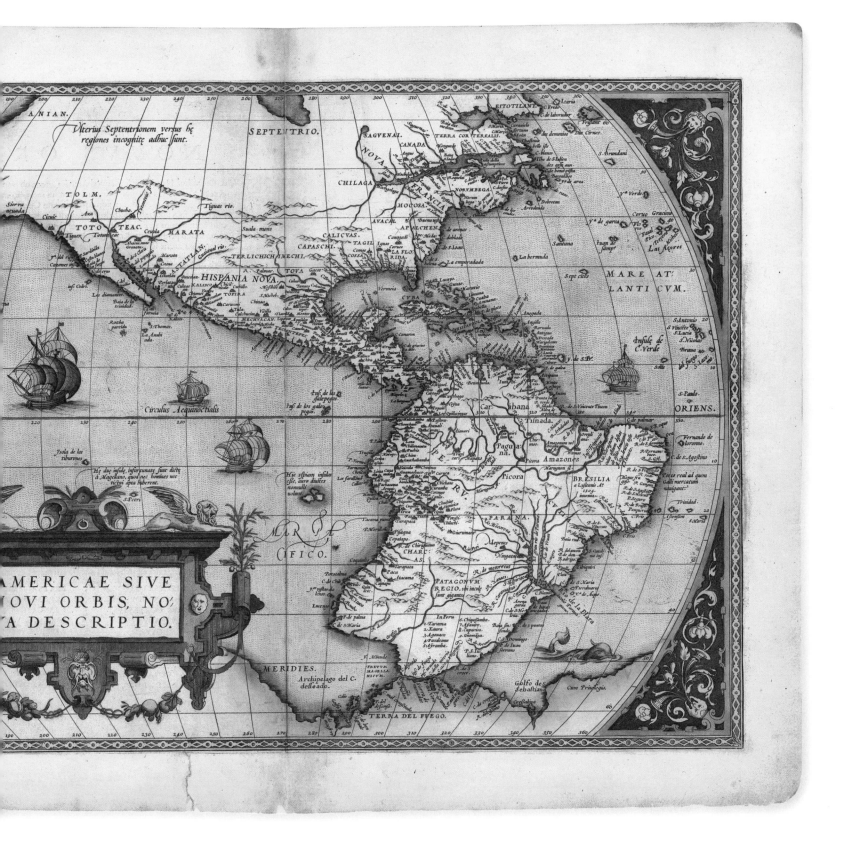

ANIAN.

Vlterius Septentrionem versus hę regiones incognitę adhuc sunt.

SEPTENTRIO.

SAGVENAI.

NOVA

CANADA

CHILAGA

ESTOTILANT.

TERRA CORTEREALIS

TOLM.

MOCOSA.

FRANCIA

NORVMBEGA

S. Brandani

Y⁴ Verde

Corvo Graciosa

Las Açores

TOTO TEAC.

MARATA

CALICVAS.

APALCHEN.

Y⁴ de garça

Sierra neuuada

Suala mons

CAPASCHI.

TAGIL

LA FLO-RIDA.

La bermuda

Sept cites

MARE AT-LANTICVM.

TERLICHICH IMECHI

TOVA

HISPANIA NOVA.

CVBA

Santana

Iuan samp

Insulę de C. Verde

S. Antonio
S. Vincēte
S. Lucia
S. Nicolas
Braua

ORIENS.

S. Paulo

MECHVACAN.

Roatha parcida

La Amuli ada

Circulus Aequinoctialis

Inf. de las galapegos

Inf. de los galapegos

Car-ibana

Tilnada.

Amazones

Picora

PER V.

Pagua-na.

Maragnon fl.

BRESILIA
a Lusitanis An.
1504.
inuenta.

Fernando de lorunno.

de S. Agostino

C. de palmar

Ysola de los tiburnes

Hę duę insulæ, infortunatę sunt dictę à Magellano, quod nec homines nec victui apta haberent.

Hie yspian insulas esse, auro diuites nonnulli volunt.

S. Petri

MAR DEL

PACIFICO.

PARANA.

Mepenes

CHARC-AS.

PATAGONVM
REGIO. *ubi incolæ*
sunt gigantes.

Trinidad.

AMERICAE SIVE
NOVI ORBIS, NO-
VA DESCRIPTIO.

In Peru.
1. Tarama
2. Kaura
3. Agonaco
4. Acalama
5. Ayramba.

6. Chiquissamba.
7. Agaure.
8. Cupacica.
9. Guanbga.

MERIDIES.

Archipelago del C. desseado.

FRETVM
MAGELLA-
NICVM.

Golfo de S. Sebastian.

Cum Priuilegio.

TERRA DEL FVEGO.

53

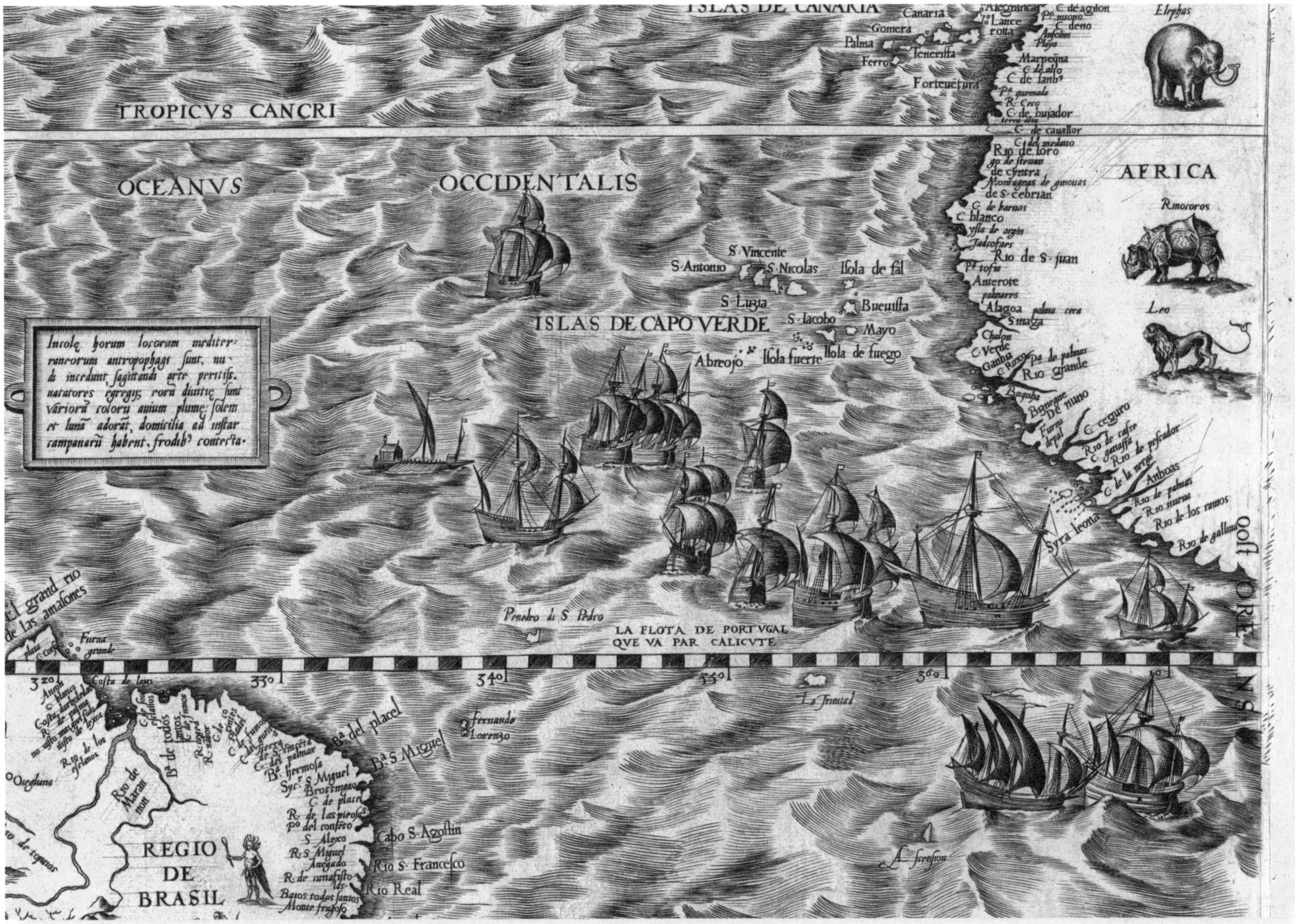
ISLAS DE CANARIA
TROPICVS CANCRI
OCEANVS
OCCIDENTALIS
ISLAS DE CAPO VERDE
AFRICA
Elephas
Rinocoros
Leo
Incolę horum locorum mediterraneorum antropophagi sunt, nudi incedunt sagittandi arte peritiss. natatores egregij eorũ diuitię sunt variorũ colorũ auium plumę: solem et lunã adorãt, domicilia ad instar campanarũ habent, frodib' contecta.
LA FLOTA DE PORTVGAL QVE VA PAR CALICVTE
REGIO DE BRASIL
Cabo S. Agostin
Rio S. Francesco
Rio Real
Syra leona
S. Antonio
S. Vincente
S. Nicolas
Isola de sal
S. Lucia
S. Iacobo
Bueuista
Mayo
Isola fuerte
Isola de fuego
Canaria
Gomera
Palma
Ferro
Teneriffa
Forteuētura
Rio grande
Fernando Lorenzo

AMERICAE SIVE QVARTAE . . . A similar declaration of supremacy can be found on *Americae Sive Qvartae Orbis Partis Nova et Exactissima Descriptio. . .* (1562), by Hieronymus Cock of Antwerp (c. 1510–70). In recognition of Portugal's power, one of its trading fleets sits off the west coast of Africa (enlarged, *above*) massing for its journey around the Cape of Good Hope to India. Cock's map is based on the charts of the Spanish cartographer Diego Gutiérrez (c. 1485–1554). Gutiérrez collaborated with British navigator Sebastian Cabot (c. 1476–1577) and succeeded him as pilot major of the Casa de Contratación in Seville—the central trade agency of the Spanish Empire, which also functioned as the chief clearinghouse for navigational and cartographic information about the New World. This also was the largest map printed to date—33 by 34 inches, on a 39-by-40-inch sheet—and the first to name California.

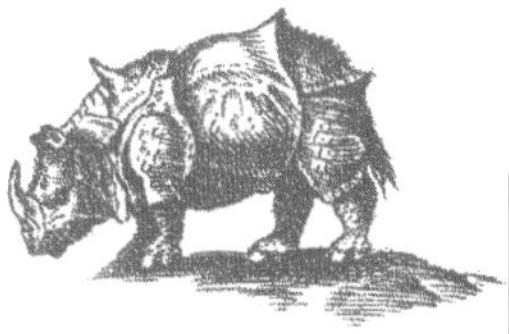

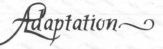

Adaptation

Because many of the decorative images on maps, including ships, were generic, and because there were no copyright laws protecting a cartographer's imagery until the eighteenth century—in Great Britain several Engraving Copyright Acts were passed beginning in the 1730s—such images were regularly borrowed. Consequently, before the 1730s, similar-looking ships appeared on different maps and on different oceans on various maps over time. In fact, by the mid-sixteenth century, in Ortelius's home country of Flanders and elsewhere, engravers and printers were producing sheets of ornamental prints of ships, cartouches, beasts, and the like—precursors of modern decals and clip art—for purchase and use by other engravers in decorating their maps and other graphics. Ortelius certainly utilized these products liberally in the *Theatrum*. He and others adapted and personalized these ornamentations for their own uses.[12] Today such personalized touches often help scholars identify maps of unknown provenance.

Cartographic images of ships serve as essential historical sources for naval architecture and activities, including fishing, commerce, and warfare. The further back in time one looks, the fewer are the sources of technical information on ships and ship-building. Literary or graphic, those records that have survived usually provide mostly conventional information. Especially with regard to old ships, there are very few remaining plans or sets of written directions for building them. That lack of sources presented a particular set of problems during preparations for the quincentennial celebration of Christopher Columbus's voyage of discovery to the New World in 1492. As part of the commemoration, two replica fleets of Columbus's three ships—the *Niña*, the *Pinta*, and the *Santa Maria*—were reconstructed to sail once again across the Atlantic. The *Niña* and the *Pinta* were said to be caravels, and the *Santa Maria* was probably a *nao*, an older, more general Spanish designation applied to somewhat larger ships during Columbus's time. But no period plans of caravels or naos existed to guide the reconstructions. Contemporary pictures of the three ships—including those on the eight wood-cut maps and other illustrations in the 1493 letter account of the voyage that Columbus published in Basel, Switzerland—were, for the most part, largely allegorical and inaccurate;[13] those illustrations, non-specific and based on contemporary models, were drawn by artists who had perhaps read Columbus's descriptions but had not seen the New World or the fleet. As a result, those planning the 1992 quincentennial had to draw from a broad spectrum of expertise and mine numerous sources to complete the ship-building projects. Many late-fifteenth- and early-sixteenth-century reports and maps illustrating ships, such as those by Ortelius, Münster, and others, were painstakingly reviewed to guide the construction and make the replicas as historically precise as possible.

Woodcut illustrations from De Insulis Nuper Inventis: The Letter of Christopher Columbus Announcing the Discovery of the New World, *1493.*

Fishing, Commerce, and Warfare

In a similar manner, by studying old maps and by reading journals and other reports, historians can gain insights into nautical activities of the more distant past. An excellent source of historical nautical maps is the *Vallard Atlas* of 1547. Not a true atlas, as the charts are largely unrelated and of different styles, sizes, and scales, it is composed of fifteen magnificent untitled Renaissance portolan chart–style manuscript maps of different areas of the world. (A manuscript map is hand-drawn and unique, as opposed to one printed in multiple copies.) The *Vallard Atlas* is a product of cartographers from the Dieppe school, the name given to a group of chart makers of diverse backgrounds that sprang up in and around the major Renaissance port of Dieppe, on the French coast of northern Normandy. Dieppe was an important center for global trade (in, for example, fish, dyewood, and textiles) and exploration (of the East Indies and other locales). The Dieppe school of cartography flourished under the reign and patronage of King Francis I (1515–47).

The authors of the charts of the *Vallard Atlas* are unknown to us today, but the charts are most likely the works of a Portuguese cartographer and at least one other chart maker and are possibly modeled on Portuguese maps. The atlas is likely named for its first owner: "Nicolas Vallard de Dieppe, 1547" is written on the first page, and what might be his coat of arms appears on a chart of northeastern Brazil. While Vallard may have been a chart maker of Dieppe, he does not seem to be the author of any of the charts in the atlas.

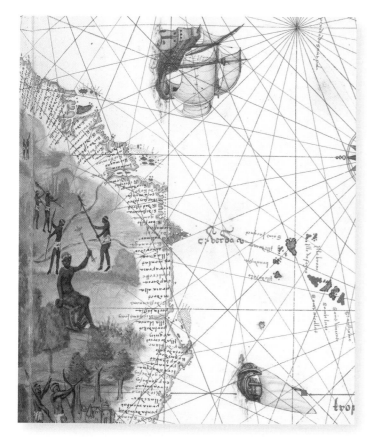

VALLARD ATLAS, CHART 7, NORTHWEST AFRICA, SHIPS. The Vallard charts include many beautiful, accurately rendered ships working the seas of the world in different pursuits. On the seventh chart (see full chart on page 139) there are two such vessels off the coast of northwestern Africa; in all probability they are Portuguese and engaged in exploration and trade. To the south, a perfectly drawn caravel is sailing northward; to the north of it is a smaller ship powered only by a characteristically diagonally mounted lateen sail.

VALLARD ATLAS, CHART 10, WEST INDIES, MEXICO, CENTRAL AMERICA, NORTHERN SOUTH AMERICA, CARAVEL. On the tenth chart, far to the east of Florida, a caravel is unusually and stunningly shown from the rear, underway with its main sails fully expanded. Many of the Vallard ships are shown sailing with the wind (expanded sails) or around it (flat sails). Awestruck natives onshore sometimes observe the vessels. Other ships are shown near large sea creatures, to indicate the vigorous pursuit of fish and whale by-products, always a valuable cargo. All such depictions are European testaments to European seamanship.

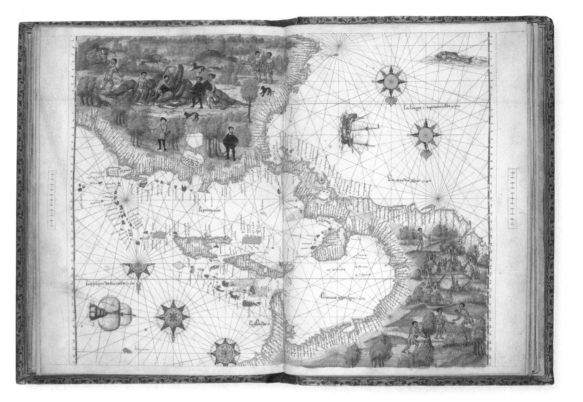

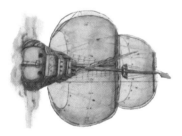

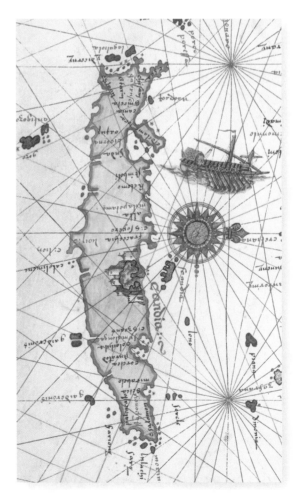

VALLARD ATLAS, CHART 15, AEGEAN SEA, GALLEY. On the last two charts of the atlas, galleys, many from Venice and carrying out the commerce of that empire, correctly ply the Adriatic and Aegean seas, respectively (some maps incorrectly show galleys on Atlantic and Pacific ocean voyages); a particularly nice one, propelled by its oars alone, appears off the northern coast of the island of "Candia" (Crete), seen here.

IRLANDIAE ACCVRATA DESCRIPTIO, WHALING. Fishing and the commerce in fish are among the nautical activities continuously and explicitly depicted on charts well beyond the *Vallard Atlas*. Boazio's *Irlandiae Accvrata Descriptio* (c. 1606) (see full map on page 24) contains an early depiction of whaling. At the bottom, just to the left of center, is a large sea creature that looks like a big fish but is undoubtedly meant to be a whale (see enlargement, *above*). To the left of it is a ship emanating billows of smoke from the onboard rendering of a whale for its precious oil; or perhaps the ship, as sometimes happened, accidentally caught fire during the reduction process. Just to the right of center at the top of this map, a man in a small boat wields a trident-like harpoon (*left*).

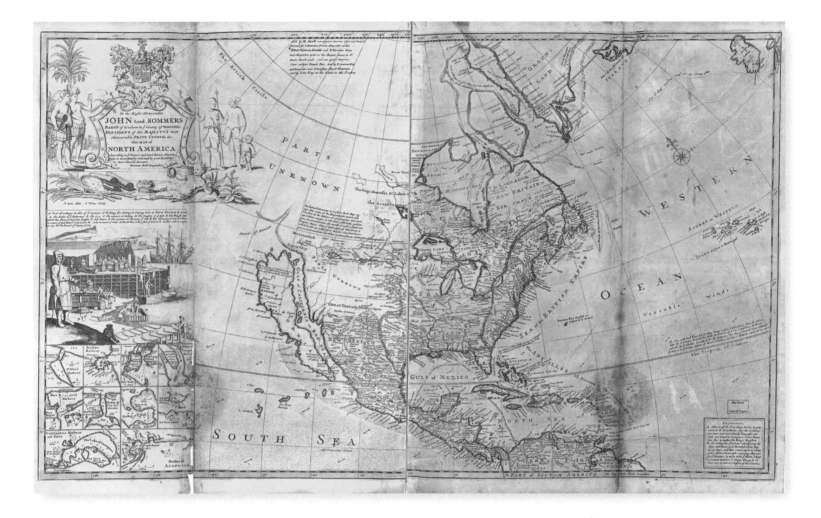

NORTH AMERICA. In a detailed inset (enlarged, *opposite*) in the center on the left side of Moll's map of North America (c. 1717), known as the Codfish Map, a three-master ship on the northeast coast of Newfoundland takes on a cargo of codfish caught on the Grand Banks. On the land, Europeans and their native partners not only display the tackle for catching cod but busily process it as well. They are cleaning, salting, and drying it on tables; once it is dried, they store it on shelves for shipment. Below this inset are others of maps indicating where along the coast some of these activities took place. Above the cod-fishing industry inset is the map's cartouche.

On its right side are natives dressed for winter, holding harpoons used to hunt whales and seals. Cod from the North Atlantic was to the French and British what the gold and silver treasures of the Aztecs and Incas had been to the Spanish in the sixteenth century. Moll's map is particularly important for showing European and American audiences these disputed fishing grounds, which were a factor in the colonial wars fought between the French and British in North America from 1689 to 1763. Like Moll's Beaver Map, the Codfish Map too could be purchased either as a single map or as part of his great atlas, *The World Described . . . .*

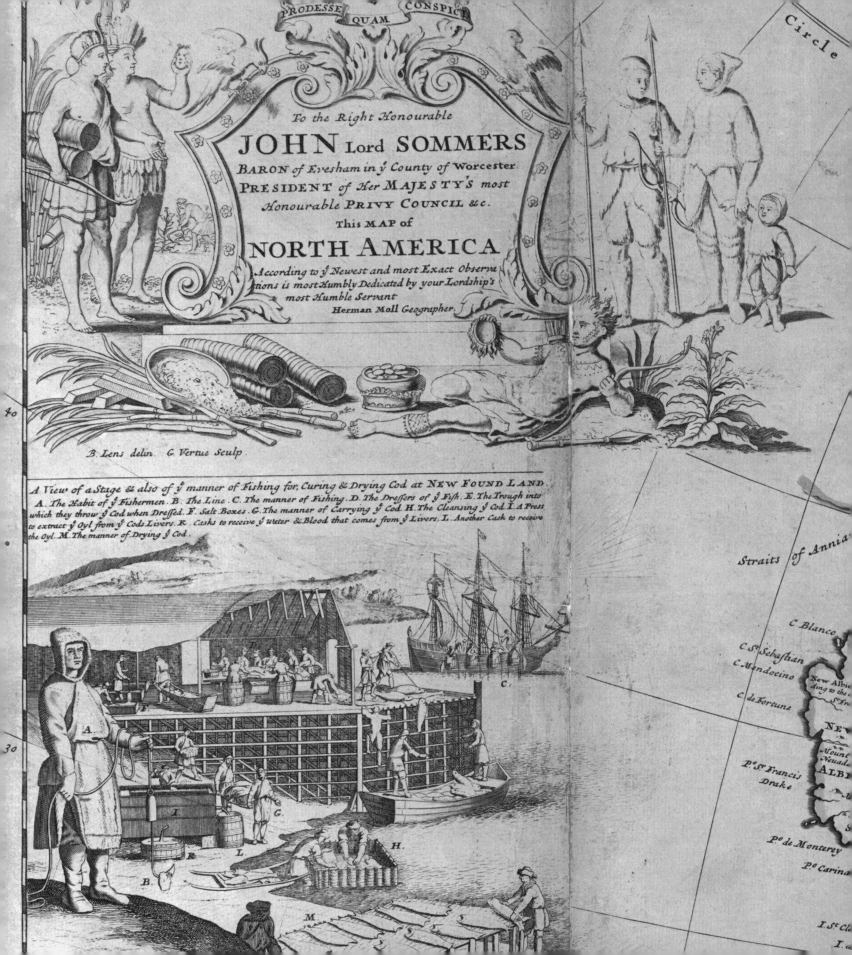

PRODESSE QUAM CONSPICI

To the Right Honourable

JOHN Lord SOMMERS

BARON of Evesham in y County of Worcester
PRESIDENT of Her MAJESTY'S most
Honourable PRIVY COUNCIL &c.

This MAP of

NORTH AMERICA

According to y Newest and most Exact Observa
tions is most Humbly Dedicated by your Lordship's
most Humble Servant
Herman Moll Geographer.

B. Lens delin. G. Vertue Sculp.

A View of a Stage & also of y manner of Fishing for, Curing & Drying Cod at NEW FOUND LAND.
A. The Habit of y Fishermen. B. The Line. C. The manner of Fishing. D. The Dressers of y Fish. E. The Trough into
which they throw y Cod when Dressed. F. Salt Boxes. G. The manner of Carrying y Cod. H. The Cleansing y Cod. I. A Press
to extract y Oyl from y Cods Livers. K. Casks to receive y water & Blood that comes from y Livers. L. Another Cask to receive
the Oyl. M. The manner of Drying y Cod.

Circle

Straits of Annia

C. Blanco

C. St Sebastian
C. Mendocino

C. de Fortune

New Albi
ding to the St Fr

NE

P.t St Francis
Drake

Mount
Neuada
ALB

P.o de Monterey

P.o Carina

I. St C

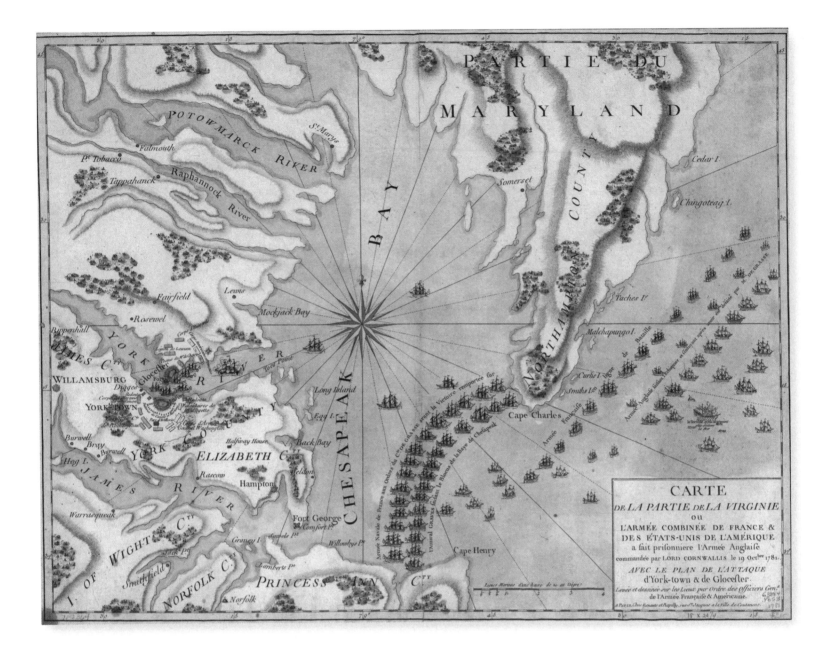

CARTE DE LA PARTIE DE LA VIRGINIE... Especially from the late seventeenth through the early nineteenth centuries, images of combatant ships on maps commonly depicted warfare at sea. Employing more than one hundred period ship images, *Carte de la Partie de la Virginie . . .* , published by Esnauts et Rapilly in Paris in 1781, vividly portrays the naval might of the combined American and French fleet in the Chesapeake Bay; French naval power helped lead the United States to victory over the British during the Battle of Yorktown and ended the Revolutionary War. The ships are in combat, blockade, and bombardment deployments.

BIRDS EYE VIEW OF LOUISIANA . . . In the tenor of Robert Herrick's quotation at the beginning of this chapter, two more fine examples of ship illustrations are found on the popular cartography of the American Civil War. On the lithographed *Birds Eye View of Louisiana, Mississippi, Alabama, and Part of Florida*, published in New York in 1861, from John Bachmann's popular Panorama of the Seat of War series of maps, contemporary ships of various types under full sail as well as ships powered by a combination of sail and steam demonstrate the massive and highly successful Union blockade of New Orleans and its Gulf Coast environs that had begun in April 1861. Even viewers who were not adept at reading more involved maps could understand its message. By the end of April 1862, New Orleans had surrendered to Union forces.

FOLLOWING PAGES: THE SIEGE OF YORKTOWN, APRIL 1862 is another bird's-eye-view lithograph of the Civil War. It was drawn by C. Worret, an actual observer of the battle of Yorktown, and published in Baltimore. Union ships and ironclad gunboats, many of them steam-powered, are bombarding the historic and strategic Confederate Virginia city at the confluence of the York River with the Chesapeake Bay. Surrounded by Union land and sea forces, this important battle, part of McClellan's Peninsular Campaign, ended on May 4 with the capture of the city.

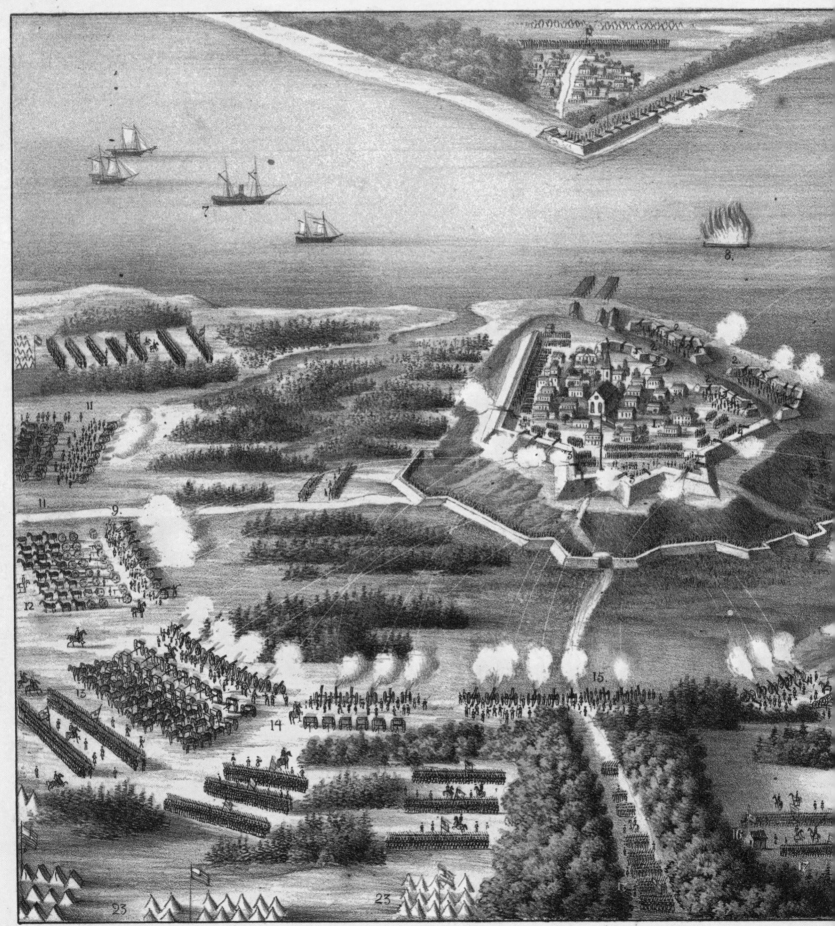

Publ. by C.Bohn 268.Penn.Av.Washington,D.C. and Old Point Comfort,Va.

Ent. accord. to Act of Congress A.D. 1862 by C.B

Drawn on the Spot by C. Worret 20. N.Y.R.

Arnarhord

Glama.

...VNG

Perpetuæ
nives.

Talknafiord.

Patrix fiord.

Kolreug

Bredevig.

Raugu
Sand

Hualo ruo

Occidens

Bard

Barda
Strang

Vadil

Flatey.

Bin
eyar

Perue Strand

Huams Suert.

Fons
quar
auan
mutu

Huams
fiord

C

Breydafiordur

Altafiord

Hiatus terræ
fœtentes

Galnla
vig

Hrumfiord

Kolgurfiord

Grimsta
fiord

Helga
velt

Kumbrum
vig

Skegur Srrand

Seyra Sneit

Stromfiord

Borgerfiord

Melasta

rhallaryfiord

Brimnes

D

Ondvertnes

Sneuels Iokul.

Akranes

Hualf

Staps

Stadarsted.

Hersey

Videy
closter

Skerifiord

Londranga

Hellur

Haffiorderey

Hasnarfiord

G

E

Rarmalanes

Slavig

Rykianes

Fost

commutun
nas nigras u
albas

Eldey

Tangi

F

Geie fuelasker

Geie eiar

H

3

Denizens of the Deep

"Yea, slimy things did crawl with legs
Upon the slimy sea.
About, about, in reel and rout
The death-fires danced at night;
The water, like a witch's oils,
burnt green, and blue and white."

—Samuel Taylor Coleridge,
from Rime of the Ancient Mariner, 1798 [14]

wellers on the world's land-masses have only gradually, over the last thousand years or so, come to see the waters that make up the vast majority of the planet as "highways" rather than barriers. Until the era of the Vikings of the West, the oceans were universally thought to be frightening abysses, fraught with unknowns, mysteries, and dangers; some people in the 1400s still held that the seas boiled at the equator and some in the 1700s that sea monsters prowled the depths. Medieval peoples who perceived the earth to be flat believed that the vast oceans stretched to the ends of the world—indeed to its very edges.

The seas were the sources of great miasmas and raised terrible and devastating storms that sometimes wreaked havoc on land. They also were inhabited by large numbers of strange and conceivably treacherous creatures, some of which came to be of increasing value to people ashore. These species included herring, which swarmed in immense, silvery shoals; cod, angry-looking fish that can grow to almost the size of a man; and whales, leviathans of the sea. The open seas could be somewhat, although not completely, trusted—only from the land to the horizon or, when sailing, until those on a boat lost the sight of land.

From the eighth century to the tenth century, the Viking Norsemen ruled the seas surrounding Europe. They were among the first to set forth across the open waters, in unique longships, and Viking voyages of trade, exploration, colonization, and plunder reached out as far as the Western Hemisphere, Russia, and southwestern Asia. And they did so without maps, other than mental ones, basing their navigations on such natural phenomena as the stars and on rudimentary sun compasses, landmarks, and knowledge passed down from generation to generation.

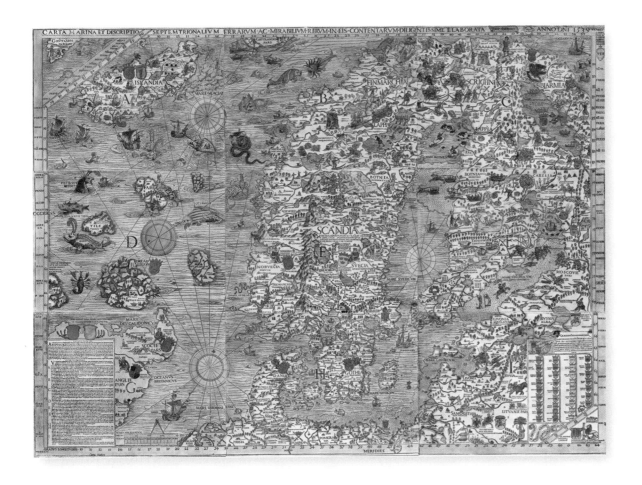

CARTA MARINA. Olaus Magnus, a Swedish writer and Catholic ecclesiastic, created this amazingly detailed map to accompany his treatise on Scandinavia entitled *Carta Marina et Descriptio Septemtrionalium Terrarum*, published in Venice in 1539. The first mostly accurate map of Scandinavia, it shows seas abounding with fantastical creatures of the deep.

After the waning of the Viking Age, sailing on the open seas increased; by the later Middle Ages—with the revitalization and expansion of urban centers and regional commerce, and symbiotic advancements in marine skills, science, and technology, including cartography—seafaring began to truly thrive. Along with a greater command of the oceans came a new perspective on the potential benefits of exploring the realms that lay across the waters. In his 1713 poem "Windsor-Forest," British poet Alexander Pope (1688–1774) envisioned that "the time shall come, when . . . seas but join the regions they divide." The growth of seafaring in turn stimulated the development of cartography, and navigators demanded more and better maps. Cartographers obliged, issuing maps embellished with representations not only of the ships that came to ply the seas, but also of the newly encountered actual and fantastical denizens that inhabited the alien deeps below.

PREVIOUS PAGES: *Right*, **ORBIS TERRARUM NOVA ET ACCURATISSIMA TABULA.** Nicolaes Visscher, Amsterdam, c. 1690. *Left*, **ISLANDIA,** from *Theatrum Orbis Terrarum*. Abraham Ortelius, Antwerp, 1590.

Extraordinary Creatures

Early cartographers worked in their home countries, far from those who sailed on the seas. They largely had their own experiences and imaginations—or those of fellow mapmakers—to rely on as sources, and thus they freely interpreted navigators' descriptions of encountered creatures. The waters on the charts of the *Vallard Atlas* of 1547 are populated by a diversity of extraordinary sea creatures.

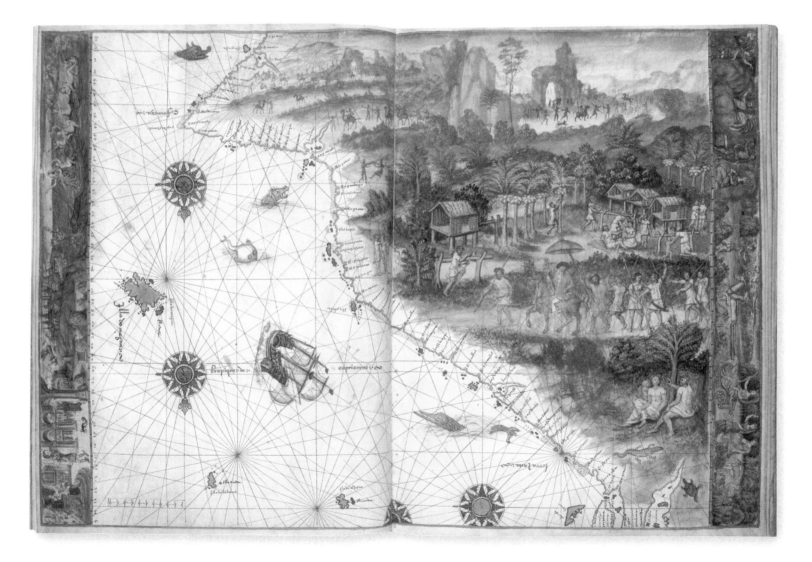

VALLARD ATLAS, CHART 1, TERRA JAVA.
On the first chart in the atlas (*opposite*), which features a part of what is labeled "Terra Java" in Australia or the East Indies, four strange "fish" are depicted. In a weird and wonderful mixing of land and sea biology, the uppermost creature in the first map of the *Vallard Atlas* seems to have the head of a wolf and the one below it the head of a wild boar (*above*). The third beast from the top could be a type of relatively common blowfish. The bottom one is some sort of sea reptile, perhaps related to the more easily recognizable crocodile, located just onshore in the lower right corner of the map near a turtle (*right*).

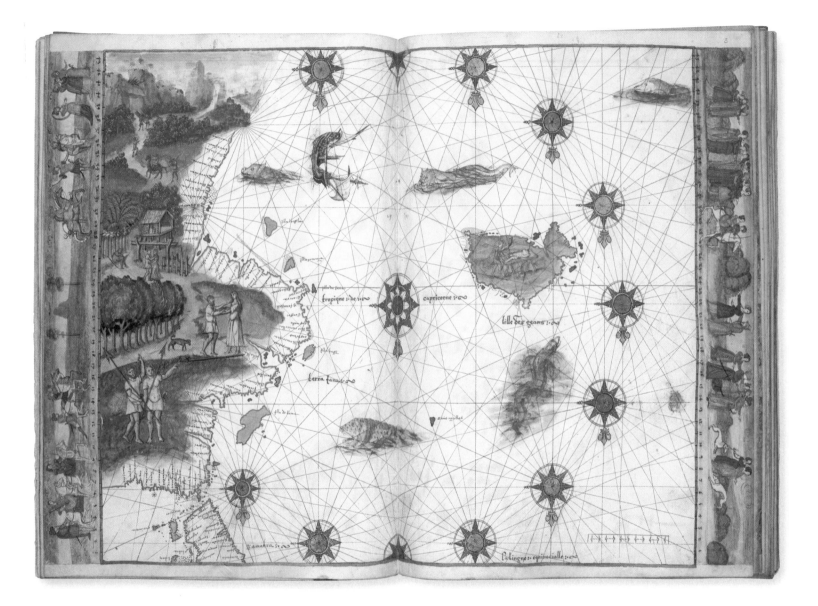

VALLARD ATLAS, CHART 3, TERRA JAVA. This chart, also of Java, shows five remarkable sea beasts. In the upper right corner is a passive, nondescript large fish, or does it have the head of a possum-like land animal? To its left is a longer, eel-like creature with the head of a lion, and toward the coast swims a fish that looks like a wild boar. Near the bottom center of the chart is what seems to be a whale, identifiable by its size, disposition, and barnacles. (Whaling had been practiced by northern coastal cultures since prehistoric times and became an outright industry in the West in the late sixteenth century. Even so, whales and whale carcasses were often mistaken for sea monsters.) To the right of the whale, farther out in the cartographic blank space of the Indian Ocean, an elongated reptile—possibly a sea snake—slithers through the water with fins protruding from its body.

SEPTEMTRIONALIUM REGIONUM...

On the 1558 map of northern Europe by Gilles Boileau de Bouillon (c. 1510–63), from the atlas *Geografia: Tavole Moderne di Geografia . . .* (Rome, c. 1575), by Antonio Lafréry (1512–77), a great sea turtle looks as if it is flying above the Norwegian Sea in the upper left corner (*right*). Boileau de Bouillon was a Flemish cartographer and publisher as well as a writer and diplomat. Unlike the somewhat exceptional Boileau, the Vallard cartographer was more familiar with land animals than those at sea.

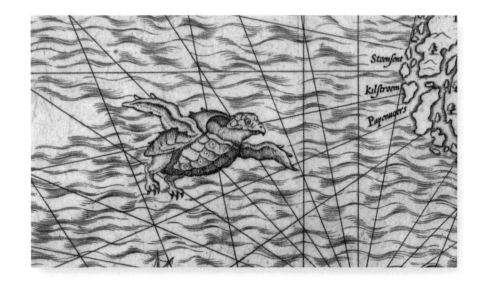

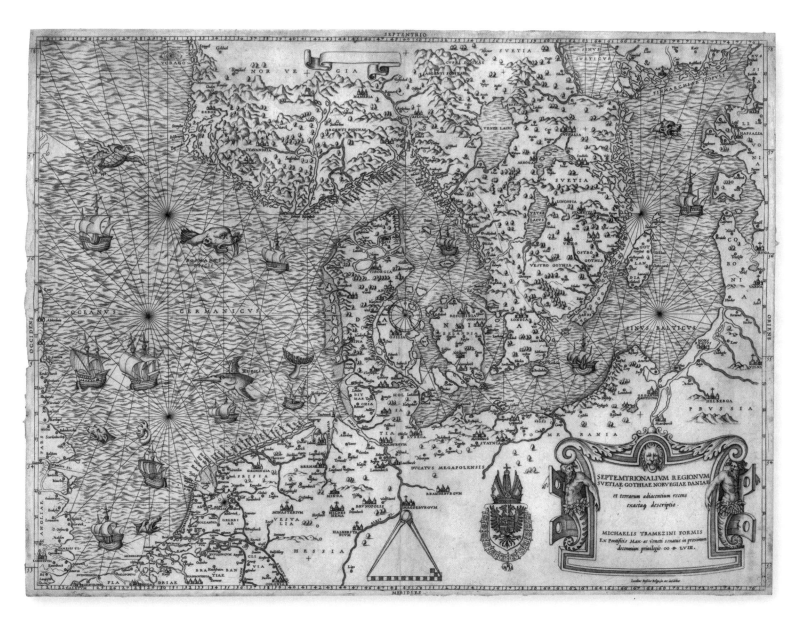

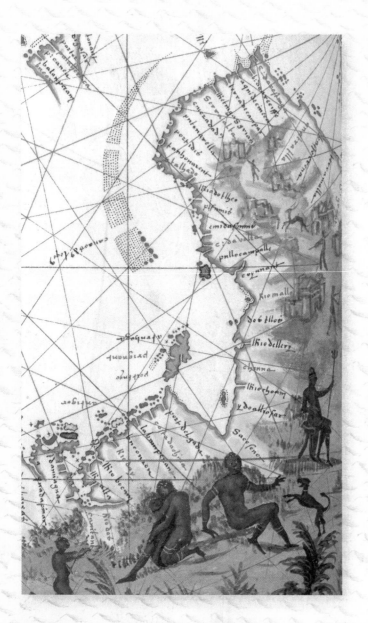

Vallard Atlas, *chart 2,* La Java, *detail.*

Cartographic Controversies

The first three charts in the *Vallard Atlas* are of Java, and they have become the most controversial of its fifteen cartographic components. In the last century and a half, a small minority of map experts have sought to prove that the coastlines shown on the first, second, and third Vallard charts are actually of eastern, northern, and western Australia, respectively, rather than of Java. They support their contentions with numerous comparisons between coastal features on the charts and those of Australia. And since these Vallard charts are supposedly based almost wholly on Portuguese information—though that supposition is a historical interpretation based on somewhat questionable evidence and leading to perhaps dubious conclusions—these specialists further assert that they are proof of the Portuguese discovery of Australia and New Zealand sometime between 1519 and 1524, well before the more widely accepted Dutch discovery in 1606.[15]

Valuable sources of historical evidence, maps are increasingly called upon to provide greater illumination of the past. Cartographic analysis his given rise to, contributed to, and clarified numerous historical controversies in addition to that surrounding the true European discovery of Australia and New Zealand. Among the more prominent ones in which maps figure as evidence are whether the Chinese undertook voyages of exploration, circumnavigation, and trade westward via the immense fleets of Admiral Zheng He between 1405 and 1433, and the issue of the exact location of Columbus's first landing site in the New World in 1492. In almost all such cases, the more radical theories rest not on hard evidence but rather on the interpretation and acceptance of the perceived accuracy of the cartographic data, especially on older maps. Maps provide information, and end users take what they need from this information and manipulate it toward their own ends.

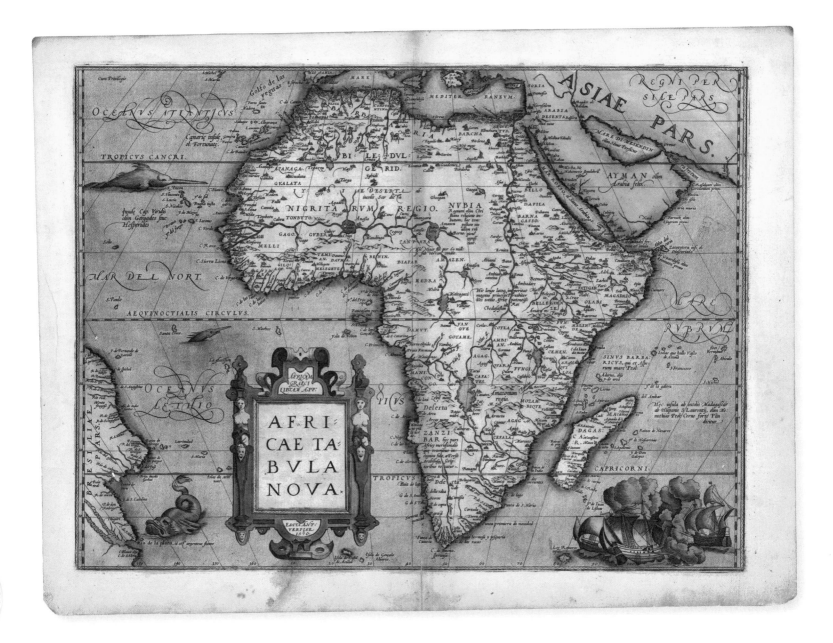

AFRICAE TABVLA NOVA. Sea denizens also decorate the myriad maps of the great printed world atlases of Abraham Ortelius and Willem Blaeu. On Ortelius's *Africae Tabvla Nova* (1570), from the *Theatrum Orbis Terrarum*, what appears to be a large swordfish is located in the upper left corner—diagonally opposite a group of ubiquitous ships. A smaller swordfish plies the waters just below the equator in the Atlantic. Just off of the coast of South America in the lower left corner swims a large fish with a serpentine, dragon-like tail.

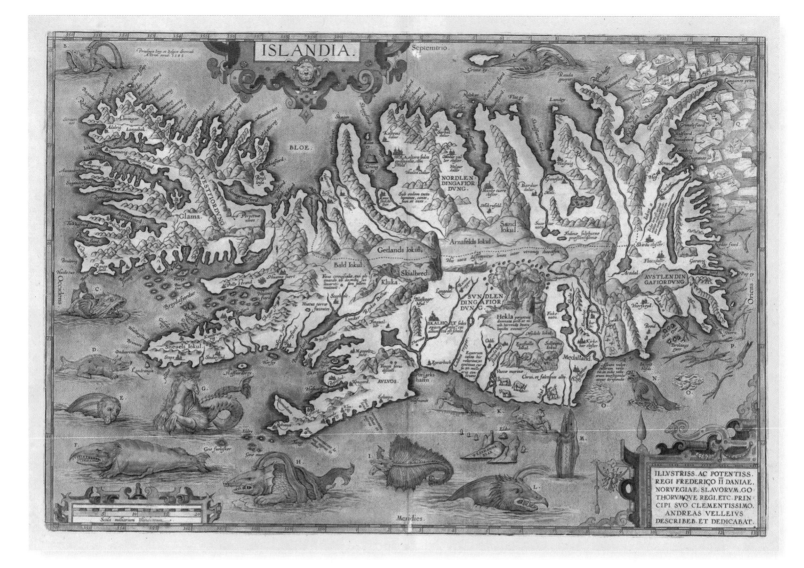

ISLANDIA. The map of Iceland, *Islandia*, first came out in the 1587 edition of the *Theatrum*. Not counting the two bulls strangely prancing in the waters off Iceland's south coast and the fifteen polar bears cavorting on the ice floes of its northeast coast, the island is encircled by no less than twelve rather ferocious-looking North Atlantic sea beasts. All of the animals, both real and phantasmagoric, and even several icebergs are identified by letters. These reference quasi-fantastic commentaries in the accompanying text of the atlas, which probably were derived from actual observations. Hence, it is claimed that "N," apparently a walrus, has been observed sleeping for half a day suspended in the water hanging by his tusks imbedded in the ice, while "M," called a *staukul*, has a penchant for posturing vertically on its tail in the water—for an entire day.[16] "M" is also noted to be quite dangerous to those who travel the seas because it eats human beings.[17]

Here Be Dragons

It has become an accepted story that cartographers once put the phrase *Hic sunt dracones* ("Here be dragons") at the ends of the known world on their maps. Untrue! While dragons or dragon-like creatures appear on many maps, this phrase does not. *Hc svnt dracones* (the Old Latin spelling) has only ever been seen on a small (diameter of 13 centimeters) engraved copper globe, the Hunt-Lenox globe (c. 1510) in the cartographic collection of the New York Public Library. The phrase is written off the coast of Southeast Asia. Moreover, although many ocean-dwelling creatures on old maps may be reminiscent of dragons, these mythical fire-breathing monsters are traditionally associated with the land rather than the sea.

The Barbarie et Biledvlgerid, Nova Descriptio *(1570) map of northwest Africa from Ortelius's* Theatrum *features what can only be called a sea dragon to the southeast of Sicily off the African coast.*

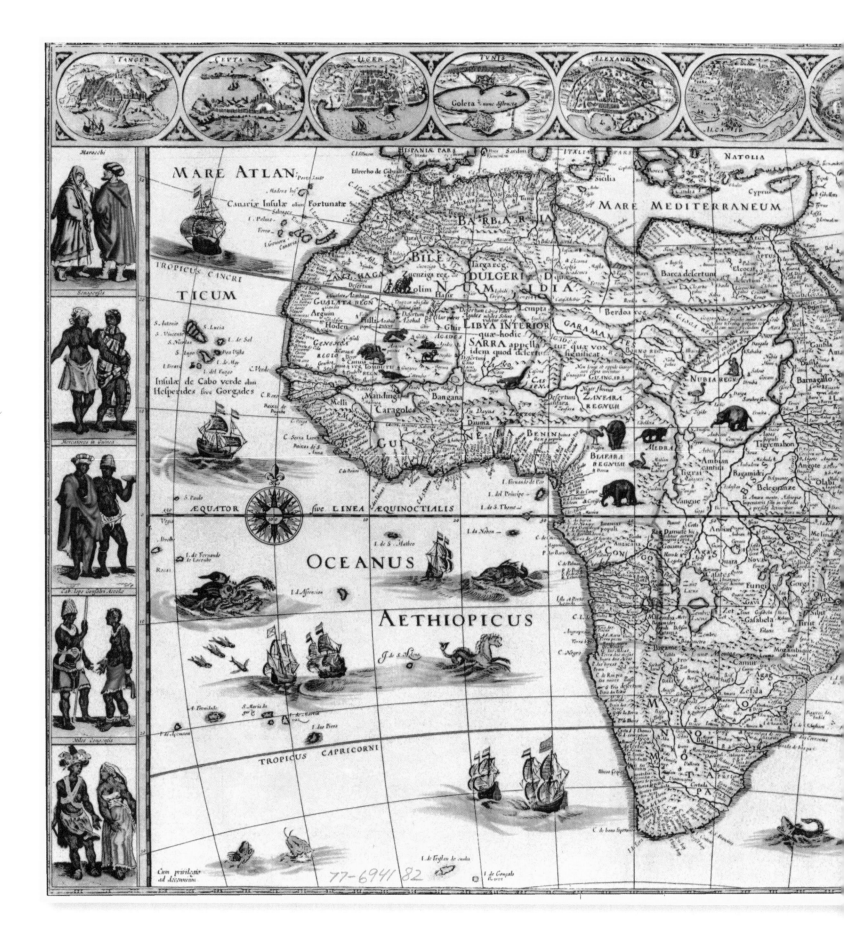

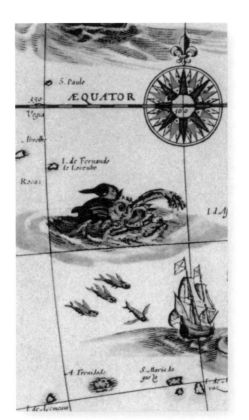

AFRICAE NOVA DESCRIPTIO. Willem Blaeu and his familial successors also published a major atlas entitled *Theatrum Orbis Terrarum, Sive Atlas Novus . . .* in 1635 and later editions. On *Africae Nova Descriptio* (1635), in the void off the continent's west coast, between the equator and the Tropic of Capricorn, are a spouting whale, a big fish, a school of flying fish, and an overly large sea horse—half horse and half curled-tail sea beast.

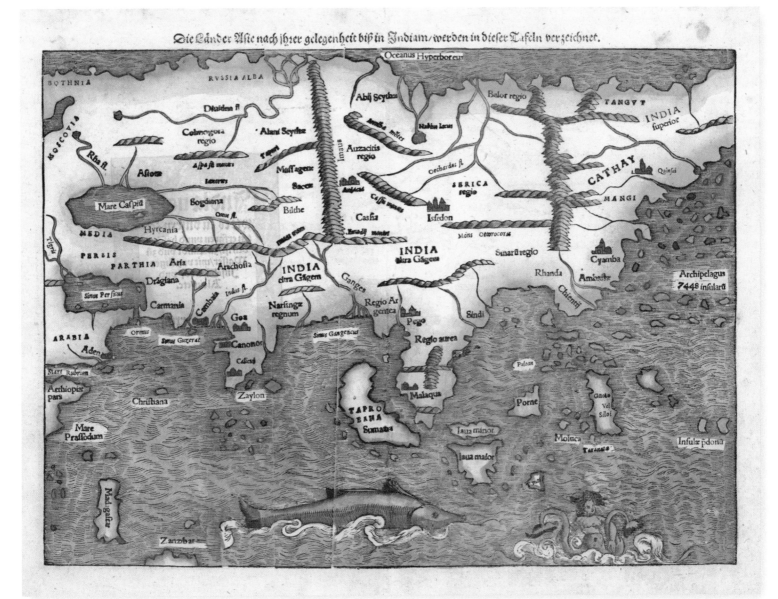

Die Länder Asie nach ihrer gelegenheit biß in Indiam/werden in dieser Tafeln verzeichnet.

VNIVERSALE DESCRITTIONE DI TVTTA LA TERRA . . . , DETAIL.
On Paolo Forlani's world map of 1578 (*opposite*; see full map on pages 12–13), two apparently friendly leviathans smile somberly in the South Atlantic off the coast of Africa; they doubtlessly represent whales. Riding on the back of the creature northeast of the ship is a beautiful and delicately engraved naked sea nymph or mermaid who holds a sail to gather the winds. The figures conceivably portend good fortune for a safe and swift passage around the Cape of Good Hope into the Indian Ocean.

INDIA EXTREMA. On *India Extrema* (*Die Lander Asie nach irer gelegenheit bisz in Indiam werden in diser Tafel verzeichnet* in German) (*above*), the map of Asia and Southeast Asia from Sebastian Münster's cartographic atlas of the four continents, *Cosmographia* (1540–1628), a great fish basks just below the island of "Taprobana" (formerly Ceylon, now Sri Lanka), which was confused at the time with Sumatra and mislabeled "Sumatra." To the right of the fish is a mermaid enjoying herself in the warm waters of the Indian Ocean. Whether hosting ships or beings, the seas on Münster's maps usually are quite peaceful, perhaps because he never sailed upon them.

NICOLAI GERBELIJ IN DESCRIPTIONEM GRAECIAE SOPHIANI. Italy-based Greek scholar Nikolaos Sophianos (active 1540s), working in Rome, compiled an eight-sheet map of Greece (c. 1544) that was published in Basel, Switzerland, by theologian Nikolaus Gerbel (1485–1560) as part of his *Descriptio Nova Totivs Graeciae . . .* (1545 and 1550). Above the cartouche, in the sea and near a ship, sits another nude mermaid on the back of a large fish, bowing a cello-type instrument. But she may not be as kindly as Forlani or Münster's nymphs. She is more a siren, warning voyagers about the dangers of seafaring. Worse, she may be trying to tempt them, like the sirens of Greek mythology whose enchanting songs were said to lure sailors to an untimely demise by shipwreck.

AMERICAE SIVE QVARTAE . . . , DETAILS. Numerous whales, large fish, smaller flying fish, sea monsters, and other creatures veritably overrun the turbulent waters surrounding the Americas on Hieronymus Cock's 1562 *Americae Sive Qvartae . . .* (see full map on page 55). Among the others are somewhat threatening mermaids offering warnings to ships passing below the Tropic of Capricorn near the Pacific access to the hazardous Straits of Magellan at the southern end of South America (*top right*). On the South Atlantic side, a kingly looking merman sits on a large fish and proclaims Spanish dominance in the area by holding a royal coat of arms. Nearby another merman draws attention to the declaration by blowing into a large sea conch horn (*center right*). In the northwest corner of the map, an angel and three cherubs display Spanish and French coats of arms. Above the Tropic of Cancer, in the northeast corner—in a further powerful allegory of Spain's supremacy—is a vignette of the king of Spain in a sea chariot drawn by sea horses, guided by Neptune, heralded by another cherub, and with a merman protecting the rear (*below*). Especially on maps of the New World, but also those of Africa, the presence of mermen and mermaids in some small part may be based on Sirenian sea cow (manatee or dugong) sightings reported by European explorers and early colonists of their tropical regions.

Flying Fish

Europeans encountered plentiful flying fish as they sailed the warmer waters off Africa and the New World, and flying fish are popular images on maps of the sixteenth and seventeenth centuries. They appear on several of Giovanni Battista Boazio's cartographic illustrations of Drake's voyage to the West Indies in 1585 and 1586, which were created for inclusion in *A Summarie and True Discourse of Sir Francis Drake's West Indian Voyage* (London, 1588–89) (see *below* and pages 85 and 86). Flying fish are also shown on a 1640 map by Jodocus Hondius (see page 87).

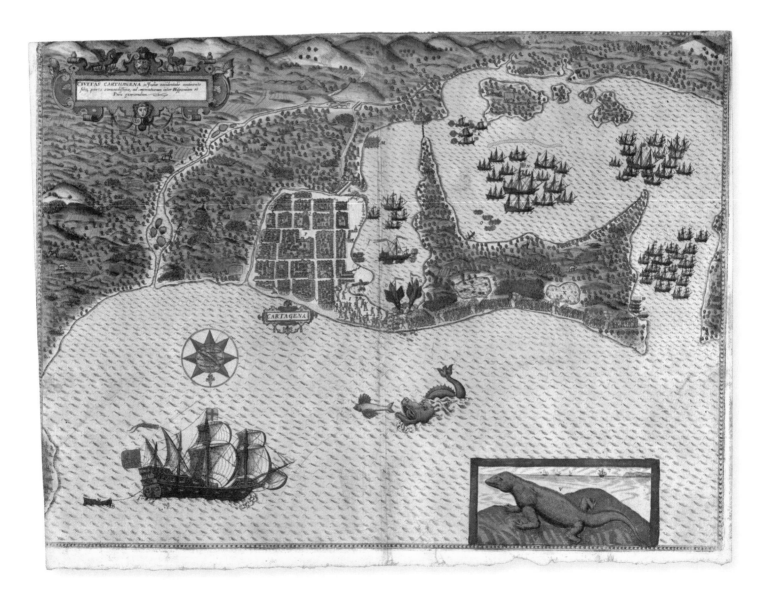

CARTAGENA. On Boazio's map of Cartagena from this series, a vignette of a flying fish about to be eaten by a sea monster fills the blank space below the port city. An iguana peacefully sitting on a rock observes the scene.

SANTIAGO, CAPE VERDE. In the harbor of Santiago, a flying fish the size of several ships occupies the lower left corner; to its far right is a spouting whale.

SANTO DOMINGO. Boazio's Santo Domingo map features a flying fish being chased out of the water by an angry sea monster who also looks to be after a large alligator. A sea turtle in the vicinity swims in the other direction. All except the sea monsters are depicted quite accurately.

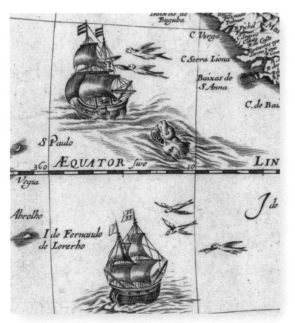

AFRICAE NOVA TABULA. Jodocus Hondius's *Africae Nova Tabula* (Amsterdam, 1640) shows birdlike flying fish soaring high into the air and almost over two Dutch ships, of seven shown on the map, off the west African coast. Some of them perhaps impressively landed on the decks of the ships, as they often did in reality. In one instance, just north of the equator, the fish are being scared out of the water by a sea monster.

Beauty, Design & Discovery

When you look at enough ocean dwellers on older maps, particularly those done prior to the Enlightenment, you notice the repetition of beings, forms, and even placement. This is due not only to cartographers and their engravers and illustrators readily and shamelessly copying from each other, but also to the functions served by these sea beasties on maps. First, they exhibited the artistry of their creators and are decorative; thus, they made the maps more attractive to prospective buyers and map users. These denizens of the deep encouraged viewers to mentally voyage to the extraordinary places depicted on the map. Simply put, appealing maps were more pleasant to see and more inviting to use.

Second, beyond contributing to the outward beauty of maps, these aquatic creatures and other adornments served as elements of design that enhanced the essential cartography. As noted earlier, these icons were sometimes used to fill in awkward and distracting blank spaces. Such voids resulted from omissions—that which was unknown or was otherwise edited out by the cartographers— that could disrupt the designs of maps and in so doing hamper their effectiveness.

White spaces also resulted from the depiction of boringly uniform large bodies of water on maps. So why not bring up beings that lurk below to pierce the surface and/or put ships on it to break up the monotony? Filling in the blank spaces, even somewhat, appreciably lessened their negative bearing on maps. In addition to being more widely used, aesthetically pleasing maps were more likely to be hung on walls or otherwise displayed.

Finally, embellishments like fish and other residents of the oceans and seas made the spectators and clients of maps feel more a part of the stories the maps had to tell. They educated map users about the diversity of sea life beyond their own shores as well as about the myths and realities that contribute to the cultural heritage of a place or region. Ornamented maps also allowed people to share in the adventures of discovery and exploration in their own times and long ago.

In the eighteenth century, with the coming of greater scientific understanding of ocean environments and scientific advances in cartography, mapmakers began to turn away from archaic, distractive embellishment in favor of cartographic content, and the denizens of the deep began to disappear from maps.

Now that we have examined some of the cartographic imagery of seafaring and of the life around the ships and their sailors, it is time to come ashore and consider the flora, fauna, and humanity found there, both familiar and exotic.

CARTA MARINA, DETAIL
(See full map on page 69.)

of the most
IMPORTANT PLANTS
WHICH ARE USED AS FOOD FOR MAN:
with indications of the
ISOTHERES & ISOCHIMENES.
OR LINES OF EQUAL SUMMER & EQUAL WINTER TEMPERATURE.
BY PROFESSOR BERGHAUS, BERLIN.

Representation
of the
FORM OF PLANTS Cultivated in the TORRID ZONE.

Tropic of Cancer

Mexico

Misantla
Vera Cruz
Colima
Tabasco
Yucatan
Central America
Omoa

The Great Antilles

Jamaica

S. Marta

Venezuela
Caracas

New Gran.
S.Fe de Bogota
Popayan

SOU

Esmeralda
Quito
Ecuad.
Guayaquil
Riobamba
Oxapa
Cuw. Condamina

Loxas
Payta
Juan de Bracamoros

AMERI

Huanuco

Lima
Peru

La Paz
Bolivia

The
Region of Cultivation
of
Cacao (Theobroma Cacao)
of the
Vanilla (Epidendrum Vanilla)
and of the
Coca (Erythroxylon Coca)
with a Survey
of the
FORESTS WHICH FURNISH THE PERUVIAN BARK
(Cinchona, Condaminia, C.Lancifolia, C.Ovatifolia C.Magnifolia &c.)

Tropic of C

Sago-tree Sagus farinifera
Date-tree Phoenix dactylifera
Sorgho Holc Sorgh
Bread fruit tree
Artocarpus incisus
Fruit
Banana
Tobacco
Plantain
Taro Colocasia esculenta
Jatropha Manihot
Yam root
Batata
Convolvulus Batatas

Oats

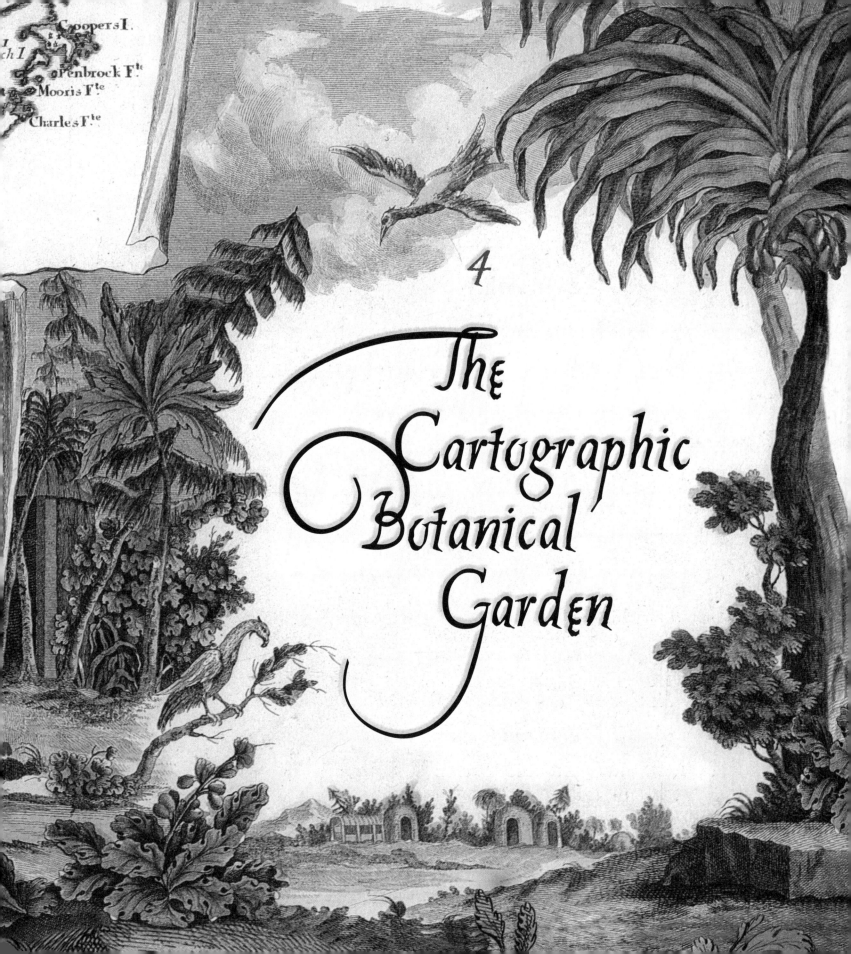

4

The Cartographic Botanical Garden

Coopers I.

Penbrock Fᵗᵉ

Mooris Fᵗᵉ

Charles Fᵗᵉ

"(Cartography) is alive with beauties —
beauties of art which appeal to the eye and to taste,
beauties of science and human endeavour
with which the whole structure is adorned and illustrated."
— Sir H. G. Fordham, Some Notable Surveyors and Map-Makers
of the Sixteenth, Seventeenth, and Eighteenth Centuries and Their Work, 1929 [18]

Into the eighteenth century, the astounding diversity of plant life from around the world rivaled animals and peoples in popularity as cartographers sought images for their maps that consumers would enjoy. When some of this flora—including potatoes and maize—began to be grown in European fields and to be served as food on European tables, interest in foreign botanicals grew exponentially.

During the great age of European exploration, this interest corresponded with an age-old quest to find the terrestrial Garden of Eden. The search for it in the Old World had not been successful, and many explorers and pioneers voyaged out with a hope of finally finding Eden—the earthly paradise.

During Christopher Columbus's third voyage to the New World, he believed he had found paradise when he came upon what is now called the Orinoco River in Venezuela, where it empties into the Gulf of Paria. He noted in his account of the voyage to the King and Queen of Spain: "There are great indications of this being the terrestrial paradise, for its site coincides with the opinion of the holy and wise theologians . . . for I have never read or heard of fresh water coming in so large a quantity, in close conjunction with the water of the sea."[19] Even those intrepid travelers who were not searching for the biblical earthly paradise often imposed the cultural metaphors of Eden on the abundant and alien landscapes and the often exotic flora and fauna of the Western Hemisphere.

Depictions of native plants on early European maps of distant places initially served a decorative function and were often generic depictions, as most of the species were as yet unfamiliar to the explorers and mapmakers. A diversity of non-specific trees dot many early maps, such as the charts of the *Vallard Atlas* (c. 1547), where such iconography both filled in blank spaces and designated vast virgin woodlands and jungles and the anticipated economic potential they represented; chart 10 of the *Vallard Atlas* (see page 58) is one such example. Northern forests especially promised not only timber, pitch, and other naval stores but also fur-bearing animals, including beavers, foxes, bears, and deer. Once cleared, these lands could serve as farmlands.

VIRGINIA . . . . On John Smith's *Virginia* (1612; see full map on pages 6–7), the area represented by almost endless woods was likely drawn based on reports by local Indians and the limited explorations and more extensive imaginations of the early English colonists. The situation had not changed much by the time of Peter Schenk's *Novi Belgii Novaeque Angliae . . .* map of 1685 (see full map on page 16), except that the huge wooded areas are now farther inland, still beyond the advancing European settlements, and populated by Indians as well as bears, deer, and other animals.

PREVIOUS PAGES: *Right*, **LE ISOLE BERMUDE.** Antonio Zatta, Venice, 1778. *Left*, **SURVEY . . . OF THE MOST IMPORTANT PLANTS WHICH ARE USED AS FOOD FOR MAN.** Heinrich Berghaus, London, 1848.

GENERALIS TOTIUS IMPERII MOSCOVITICI . . . Prior to the opening up of the transatlantic trade in furs, hides, and forest products in the seventeenth and eighteen centuries, Russia had held significant monopolies over these and related commodities for two hundred years. Russia was a major trade partner of the Hanseatic League, the powerful German Baltic–North Sea trading confederation. The natural wealth of Russia's extensive Trans-Ural Mountain hinterlands—demonstrated in the vast forests covering this c. 1704 map of Imperial Russia by Johann Baptist Homann (1663–1724)—gave her the power to control supply and dictate prices. Understandably, the up-and-coming Western European powers welcomed the growing competition from the Americas and the lower costs and somewhat easier access the New World proffered.

CARTE TRES CURIEUSE DE LA MER DU SUD...
DETAIL. The newly encountered lush jungles of the tropical regions of the world, certainly more foreign to Europeans than the Russian forests, offered similar possibilities for wealth. By the eighteenth century, cartographers began to incorporate identifiable flora and fauna on their maps—tropical forest trees and their yields of precious hardwood and dyewood, unusual fruits and vegetables, and remarkable animal skins. They also depicted native peoples, who were tragically considered as potential slaves.

Many varieties of hardwood and palm trees and other plants from the Americas and elsewhere in the partially explored South Pacific Basin crowd the southwestern corner of Franco-Dutch cartographer Henri-Abraham Châtelain's lively *Carte Tres Curieuse de la Mer du Sud...* (Amsterdam, 1719; see full map on page 23) below the Tropic of Capricorn. Their abundant yields of woods, cacao, bananas, and pineapple, among other products, are displayed in close proximity. Châtelain (1684–1743) also depicts potatoes, manioc, watermelons, indigo, and tobacco, as well as vignettes of agricultural harvesting and processing.

AMERICAE SIVE NOVI ORBIS...
(CARTOUCHE). Almost 150 years earlier than the Châtelain map, the cartouches on the various editions of Ortelius's *Americae...* (see full map on pages 52–53) were adorned with what could be Caribbean key-lime boughs and draped garlands of recognizable New World vegetables such as squash, peppers, and melons. Ortelius, working in Antwerp, a major trading center, was able to access the latest information on discoveries from around the world. The maps in his atlas are among the first to accurately show New World flora.

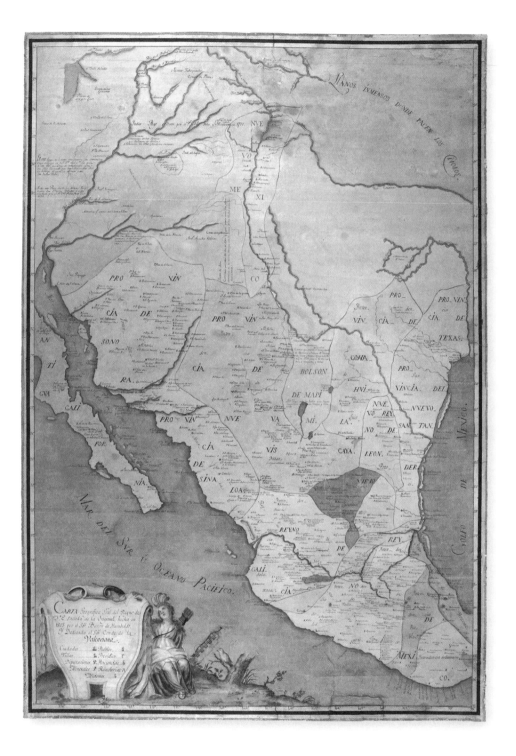

CARTA GEOGRAFICA GRAL. DEL REYNO DE N. E. . . . As in Ortelius's *Americae* (see pages 96–97), a garland of fruit, in this case bananas, is ceremoniously draped around the cartouche in the lower left corner of a rare, 1803 manuscript map of New Spain, *Carta Geografica Gral. del Reyno de N. E. Sacada de la Original hecha en 1803 por el Sor. Baron de Humboldt. Y Dedicada al Sor. Conde de la Valenciana*, attractively rendered in ink and watercolor by Juan Segura. Plants and animals on maps of the New World often function as more than just ornamental trappings. Bananas, which came from Africa with the slave trade and were grown in the tropical and subtropical Americas as food for slaves and poorer people, were traded locally and regionally. Beyond mere embellishment, the images of the bananas are a commentary on the agricultural and economic diversity of New Spain at the time. As the title of the map indicates, it is a copy made by Segura under a commission of Segura's patron, the wealthy Conde (Count) de la Valenciana of Guanajuato, after a manuscript map of New Spain by the great German naturalist, explorer, philosopher, diplomat, and author Alexander von Humboldt (1769–1859). The map was originally published as *A Map of New Spain . . . in Paris and London* in 1809 and 1810. Humboldt was a guest of the count for about a month in 1803. Segura's version of the map emphasizes Mexico's and the count's wealth (e.g., its silver mines) and its potential for wealth (e.g., its vast mountain ranges).

GVINEA PROPIA . . . The presence of palm trees and palmettos emphasized the exotic settings of the subject matter. Palms provide a fitting backdrop in an inset of a village on the West African Slave Coast (coastal areas of present-day Togo, Benin, and western Nigeria) on the map *Gvinea Propia, nec non Nigritiæ vel Terræ Nigrorvm Maxima Pars . . .*, published in Nuremberg by the Homann Heirs in 1743. The Homanns were a prominent family and firm of German cartographers from Nuremberg founded by Johann Baptist Homann (1664–1724). After his death and that of his son Johann Christoph (1703–30), the company passed to his sons-in-law and came to be known as Homann Heirs. The map of Guinea appeared in several Homann atlases over the life of the company, which finally closed up shop around 1852 after issuing more than nine hundred maps.

U MER

PROVINCE
DE
QUANG-TONG.

PROVINCE DE QUANG-TONG. The French royal geographer and cartographer Jean Baptiste Bourguignon d'Anville (1697–1782) produced more than two hundred maps in his lifetime. In the major cartouche on his map of the coastal *Province de Quang-Tong* (Paris, 1737), from his *Nouvel Atlas de Chine . . .* , there is a vignette of Chinese and Western merchants doing business (*opposite page*). The scene is flanked by two palm trees. One of the trees appears to be bearing breadfruit (sought after by a monkey); the other tree is laden with coconuts. Since China did not trade in these commodities with Europe, clearly the trees are there to decoratively create a glamorous, oriental mood for the map. While other goods were also traded, Westerners were primarily interested in obtaining silk, tea, and opium from China, for which they usually paid in "pieces of eight"—the celebrated Spanish New World silver dollar that was in wide circulation during the colonial era. During the eighteenth century, the prolific silver production of northern Mexico and the Andes helped create an early global economy. Chinese merchants showing off a silk tapestry to a European counterpart are pictured in the vignette on d'Anville's map, as is a Chinese laborer bearing a heavy chest perhaps filled with tea or opium. At the center of it all a Chinese merchant calculates the transactions on his faithful "computer," the abacus.

MAP OF THE UNITED STATES OF AMERICA. On J. H. Colton's 1850 map of the United States (see full map on page 44), the depicted plant life plays a somewhat different role. The whole image of the map, with inclusions, is surrounded by a frame largely made up of a fruit-bearing, entwined grapevine at the neatline. (The top left corner of a different printing of the map is shown here.) This type of border is not uncommon as a stylistic element on American engraved and lithographed cartography of the nineteenth century. Sometimes the edging is merely a linear rickrack; sometimes it is a statelier "banknote" design, similar to that also found on today's currency and stock and bond certificates, and is purely decorative. But more is going on with the Colton map. The encompassing fruited vine symbolizes the abundance and the opportunities, immediate and potential, of the New World and particularly the United States. In one respect, this important map offers an invitation to its viewers at home and abroad to come and join in—to help realize these optimistic promises.

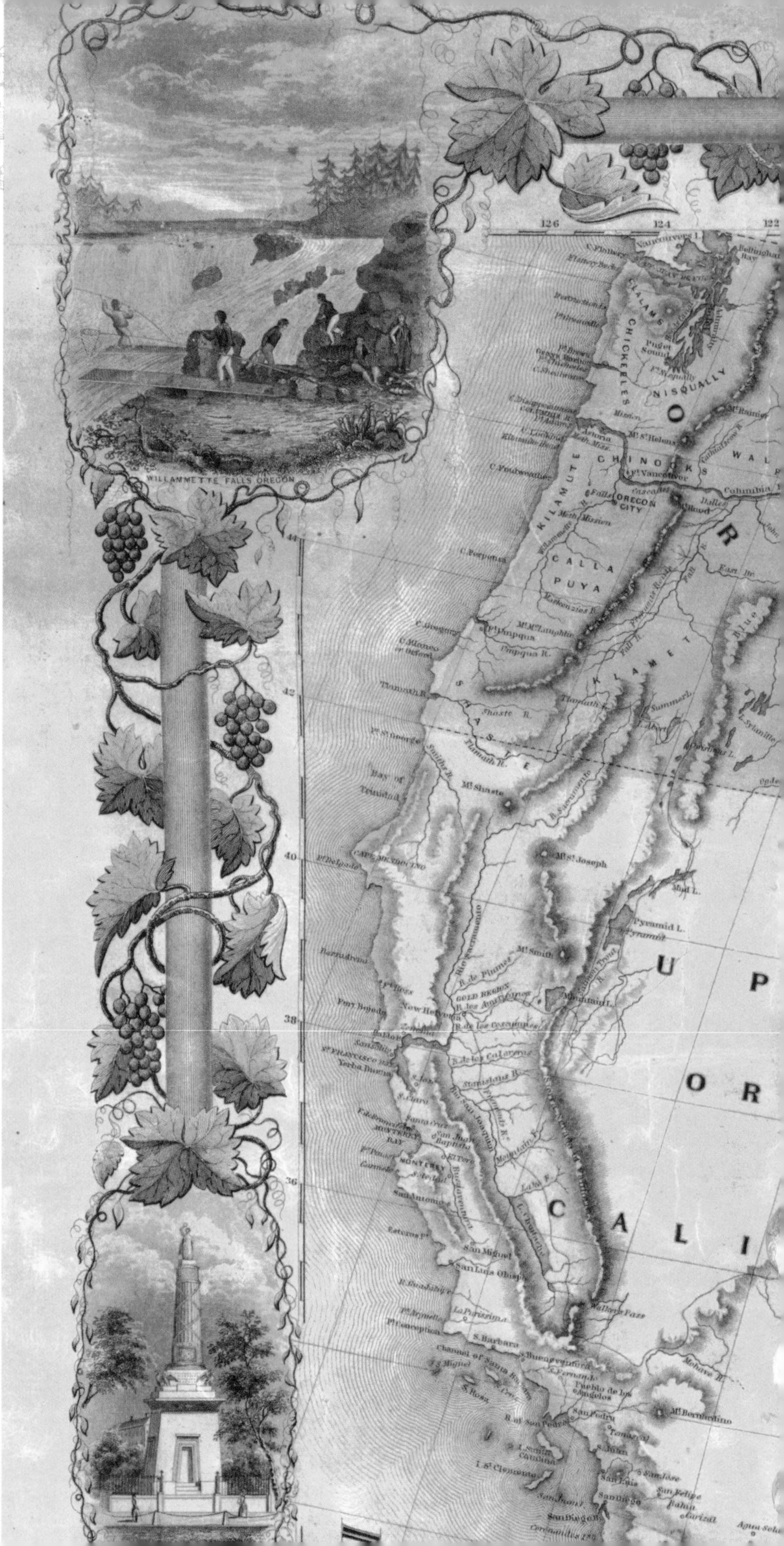

5

The Cartographic Menagerie

"Our Geographers seem to be almost as much at a loss in the Description of this north part of Scotland, as the Romans were to conquer it; and they are oblig'd to fill it up with Hills and Mountains, as they do the inner parts of Africa, with Lyons and Elephants, for want of knowing what else to place there."

— Daniel Defoe, A Tour Thro' the Whole Island of Great Britain (1724-27) [20]

The myriad animals encountered around the globe during the period of Western expansion became popular presences on many cartographic representations well into the nineteenth century. Along with images of flora, depictions of fauna were intended to give map users added insights into the regions shown, in addition to contributing to the map's attractiveness and marketability.

Animals also added to the awe and wonder of maps. For several centuries, the cartographic zoo doubtless offered its patrons access to a broader diversity of exotic specimens, actual and imagined, than real-life zoos, which, like contemporary botanical gardens, were as yet only in their infancies or adolescences. The presence of these life forms on maps and other illustrations expanded viewers' horizons and let them glimpse animal life in distant lands far beyond their own.

Animals also filled in the blank spaces of uncharted areas. Jonathan Swift, in his 1733 poem entitled "On Poetry, A Rhapsody," writes:

Geographers in Afric maps
With savage pictures fill the gaps,
And o'er uninhabitable downs,
Place elephants instead of towns.[21]

These verses and the Defoe quote that opens this chapter—literary barbs against geographers by two great writers—were likely inspired by maps such as the ones reproduced in these pages.

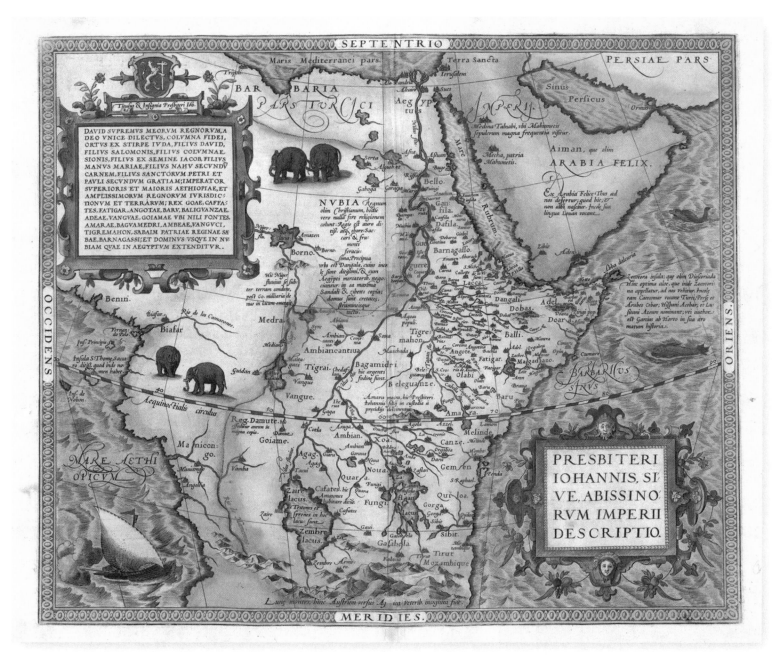

PRESBITERI JOHANNIS, SIVE, ABISSINORVM . . . Perhaps Defoe or Swift saw one of Abraham Ortelius's African maps, such as *Presbiteri Johannis, Sive, Abissinorvm Imperii Descriptio* (Antwerp, 1573–1612), from the *Theatrum.* To the west of the Nile River, as well as directly above the equator in West Africa, two elephants have clearly been strategically positioned to fill in embarrassing blank spaces—uncharted regions of the African interior. A large descriptive caption in the form of a cartouche at the upper left corner serves a similar function.

PREVIOUS PAGES: *Right*, **CANADA, LE COLONIE INGLESI CON LA LUIGIANA, E FLORIDA . . . , CARTOUCHE.** Antonio Zatta, Venice, 1778. *Left*, **LEO BELGICUS.** Baron Michael von Aitzing and Frans Hogenberg, Cologne, 1583.

Prester John

Abraham Ortelius's map on the previous page, *Presbiteri Johannis, Sive, Abissinorvm Imperii Descriptio*, depicts the alleged domains of the legendary Prester (presbyter or priest) John in Abyssinia (Ethiopia) in East Africa. The story of Prester John, who was allegedly descended from the Magi, dates back to at least the twelfth century. John was reputed to be a Christian monarch first in the far reaches of India and then in Central Asia; the basis of this part of the myth may have its roots in an actual personage who was the grandfather, foster father, or mentor of the infamous Genghis Khan. Prester John was said to be a Nestorian Christian, a member of a sect today called the Church of the East or the Assyrian Church, which was prevalent in Mongolia during the reign of Khan. Supposedly a tireless crusader on behalf of Christianity, he was believed to have vigorously defended his vastly wealthy kingdom against the forces of Islam and paganism. As the interior of Asia gradually was more fully revealed to the West, mapmakers began to change the position of the kingdom of Prester John, which eventually made the leap to Africa. Abyssinia was a reasonable location, since many of its people were Coptic Christians, and according to legend, the country was founded by a purported descendant of Solomon and Sheba—the emperor Menelik I. The belief that the kingdom of Prester John must exist somewhere in Asia or Africa held strong in Western Christian thought and Western cartography at least into the sixteenth century.[22]

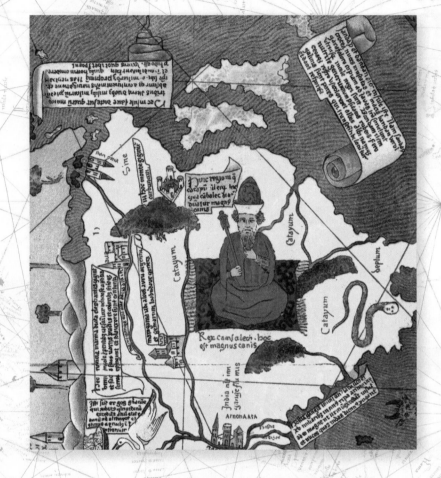

This detail of a 1457 world map by Florentine cosmographer Paolo dal Pozzo Toscanelli depicts Prester John as a ruler in India.

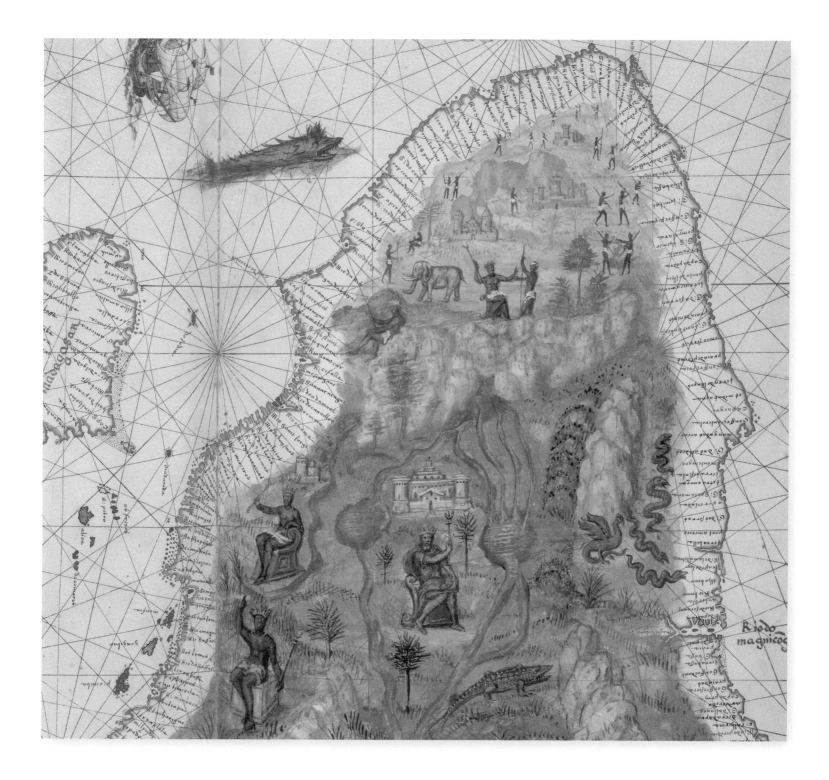

VALLARD ATLAS, CHART 5, SOUTHERN AFRICA AND SOUTHWEST INDIAN OCEAN, DETAIL.
Many of the animals on the charts of the *Vallard Atlas*, such as this one, are illustrated
so vaguely as to be almost nondescript. Those already well known to Europeans, such as
elephants, bears, and deer, are easily recognizable on the renderings of Africa and the
Americas, but those that are more exotic, including the bigger and smaller cats, elusive
rodents, and gaudy birds of the jungles and forests, are not. The latter populate the terrae
incognitae on the cartography more as bizarre generalizations than as real animals.

Buffalo and Bears

Some of the early examples of new forms of wildlife that appeared on maps were images derived by the cartographers and other artists from written descriptions, accurate or not, provided by the explorers and colonists who first observed these unknown species. The illustrations were at best reasonable secondhand graphic accounts, influenced by known beasts but not necessarily very correct, for the mapmakers rarely saw these alien creatures before they drew them. For example, when Spanish explorer Álvar Núñez Cabeza de Vaca became the first European to see wild American buffalo in present-day Texas on his expedition of 1528–36, he reported them as big woolly cattle,[23] and that is how they were initially portrayed.

TABULA TERRA NOVA, VIGNETTE. On the 1522 revised edition of Martin Waldseemüller's *Tabula Terra Nova* (see full map on pages 30–31), in the northeastern part of South America there is a large composite graphic vignette of cannibals feasting and what appears to be a female bear. (It is the bear that is important here; the cannibals will be discussed in the next chapter.) Some observers have said the animal looks more like a possum, but this cannot be, for two reasons: The creature in question has obvious external mammary glands, while the possum, a marsupial, has teats inside its marsupium, or abdominal skin pouch. And the first European report of a New World possum was by Cabeza de Vaca on his expedition of 1528–36, well after the revisions to the map—including the bear—were made.

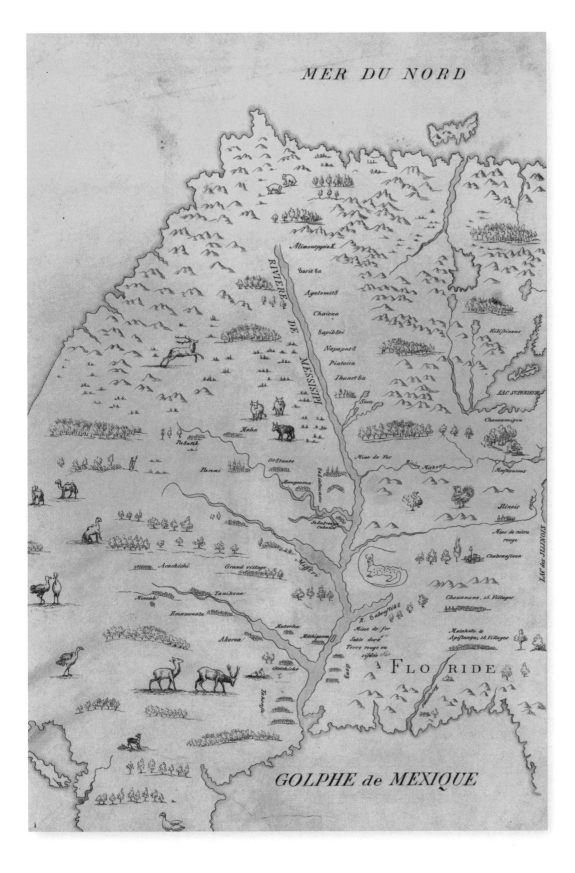

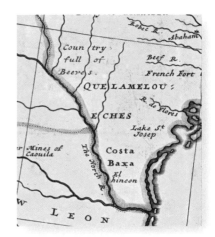

CARTE GNLLE DE LA FRANCE SEPTENTRIONALLE . . . . Several "wooly cattle" can be seen grazing to the left of the "Messisipi" River in this later English manuscript copy of the c. 1680 map by official New France cartographer Jean Baptiste Louis Franquelin, entitled *Carte Gnlle de la France Septentrionalle contenant la découverte du pays des Ilinois.* The map also has a wide variety of other animals, such as deer and a turkey, scattered about, as well as what is likely a crudely drawn alligator to the northwest of the word "Floride," and, oddly, several camels to the west.

A MAP OF THE WEST-INDIES . . . , COMMENTARY. In 1715, almost two centuries after Cabeza de Vaca first spotted buffalo in the New World, Herman Moll's map of the West Indies (see full map on pages 20–21) offered a commentary, enlarged here, along the Rio Grande on the Texas side that reads "Country full of Beeves."

The Great Beaver

Although the Dieppe cartographers were the first Europeans to represent beavers on their charts—in the *Vallard Atlas*—perhaps the most famous members of the cartographic bestiary are the beavers diligently toiling in the inset on Moll's 1715 map of British North America, *A New and Exact Map of the Dominions of the King of Great Britain on ye Continent of North America . . . .*, commonly known as the Beaver Map (see full map on page xiv and beaver detail on page xix). Although Moll's map was popular by virtue of his excellent map engraving and design skills, the beaver vignette did not originate with him.[25] In a move characteristic of mapmakers of the era, he probably copied it from a little-known 1698 wall map of the Americas by French geographer and map publisher Nicolas de Fer (1646–1720) entitled *Amerique, Divisée Selon Letendue de Ses Principales Parties . . . .* ; that image was actually a composite based on several sources, done by Nicolas Guérard, an engraver for de Fer.

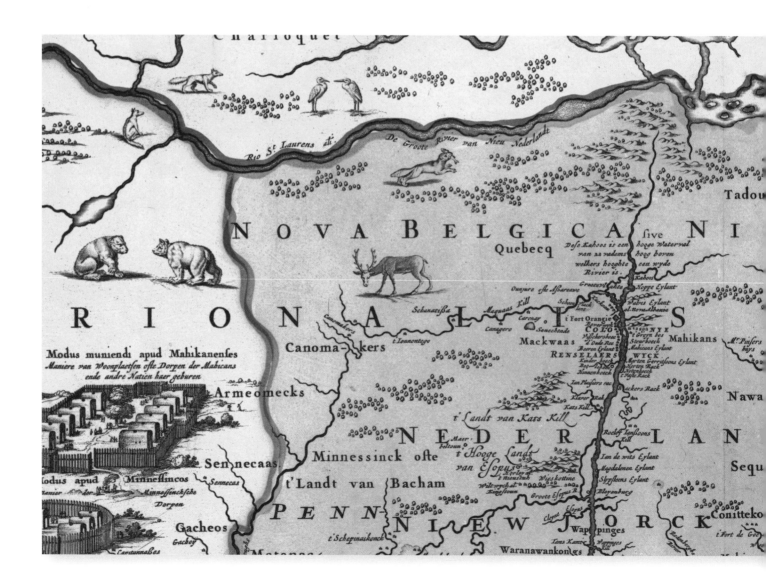

CARTE TRES CURIEUSE DE LA MER DU SUD . . . , VIGNETTE.
Henri-Abraham Châtelain probably copied the beaver vignette from Moll or de Fer for his New World–centered *Carte Tres Curieuse de la Mer du Sud* of 1719 (see full map on page 23). Abutting the beaver inset on the right at the top left center of Châtelain's map is a scene of Native Americans hunting beavers and other forest creatures and tanning their pelts for the fur trade with the Europeans.

NOVI BELGII NOVAEQUE ANGLIAE . . . , DETAIL.
On Peter Schenk's 1685 map of New England (see full map on page 16) beavers in New England and New France are depicted in more leisurely pursuits, accompanied by foxes, bears, a deer, a possum, a weasel, rabbits, herons, and a turkey, as shown in this detail. Some animals, such as the possum and the turkey, are unique to the New World, but on both maps all the creatures are clearly filling in as-yet-unexplored and unmapped areas.

113

Natural Wonders and Empire

By the time of the first publication of Moll's map in 1715, the beaver was becoming emblematic of the European North American colonies and the growing transatlantic fur trade; the craze for fashionable beaver hats led to a demand for beaver pelts. But the meanings behind Moll's decorative use of the beaver inset are conceivably more specific. He was a promoter of the rapidly expanding British Empire. While that enterprise was little more than a hundred years old in North America in 1715, for Moll the messages conveyed by the inset of cooperatively hard-working beavers at the great falls of the Niagara River were about the industry, harmony, and potential wealth and power to be realized by Great Britain in its new colony.

Another natural wonder, the Grand Canyon of the North American Southwest, was first sighted in 1540 and explored by Francisco Vázquez de Coronado's expedition of 1539–42; nevertheless, it did not appear regularly on maps until the late nineteenth century, and its full beauty was not widely recognized until the twentieth century. The Grand Canyon was generally seen as yet another geographic impediment to be overcome in exploiting a vast marginal wasteland; it was considered a "big hole" that stood in the way of progress.

That was not the case with Niagara Falls. At the same time as the Niagara River country became economically important for its furs and agriculture, as well as strategically important as a colonial frontier "shared by Britain and France" (as indicated by Moll on his map), Niagara Falls as a natural wonder became a powerful graphic and literary symbol. Thus, even Moll's anti-empire friend, the Irish author Jonathan

The Cataract of NIAGARA, some make this Water-Fall to be half a League while others reckon it no more than a hundred Fathom.

Swift (1667–1745)—who believed the resources Britain put into overseas expansion would be better spent improving conditions in Ireland—mentions Niagara Falls in his classic book *Gulliver's Travels* (1726). Gulliver, encaged by the giant Brobdingnagians in a large box, describes the sound made when an eagle drops him onto the sea as "louder to my ears than the cataract of Niagara." (Moll did the maps of the imaginary geography in this novel).[26] It is clear that, with its beaver imagery and numerous written commentaries, Moll's map served as a "booster map" to inform and lure prospective colonists.[27] It also showed its viewers the British Empire in North America and the territorial outcome of recent wars with France there, and helped counter earlier graphic claims made by such French cartographers as Guillaume Delisle.

An enlarged detail of Niagara Falls from the beaver inset on Moll's 1715 map.

The Mysterious Armadillo and Other Creatures

Another New World animal that aroused the curiosity of Europeans during the period of discovery and exploration was the armadillo. The furtive, plated creature did not hold any particular economic value for them; it was more the armadillo's peculiar build and behaviors that aroused their interest. An ancient, mostly small, generally nocturnal, and naturally armored mammal that exists in twenty species, primarily in Latin America, the armadillo is related to the sloth and the anteater. When startled, an armadillo can jump from a standstill three to four feet straight up. When threatened, some varieties can roll themselves into rigid, almost perfect balls for protection. The nine-banded armadillo now ranges from Argentina well into the southwestern United States. The armadillo remains an iconic curiosity to most people.

The first human records of encounters with the armadillo come from the Mayas of Central America—specifically, a carved relief dating back to the fourth century BCE. The first Europeans to see armadillos were the Spanish conquistadors in Central America and Mexico in the sixteenth century. Derived from their reports and those of later European interlopers, images of this strange creature soon began to appear as illustrations in accounts published by conquistadors and on maps.

LA CALIFORNIE OU NOUVELLE CAROLINE . . .

An armadillo appears in the lower left corner of Nicholas de Fer's map of California, *La Californie ou Nouvelle Caroline . . .* (Paris, 1720). It is a prominent part of the flora and fauna surrounding the partially blank cartouche, which contains the map's scale in Spanish leagues. The armadillo is on the right side of the cartouche, positioned under a nondescript shrub and facing inward. Because de Fer had never actually had seen an armadillo, his picture is slightly stylized. While the image is correct in showing distinctive bands around its middle, de Fer's armadillo also appears to have scales rather than its actual armor composed of bone and horny material. The armadillo depicted looks like a nonexistent reptilian possum.

Beneath the same bush as the armadillo, facing outward, is another generalized animal. It might be a possum or a large rodent of some sort, or even an armorless armadillo. On the left side of the cartouche are two birds. The larger is perhaps a type of heron, and the smaller may be a bald eagle; it is hard to be absolutely sure, for they too are somewhat indiscriminately represented because of de Fer's lack knowledge; the images are primarily decorative in function.

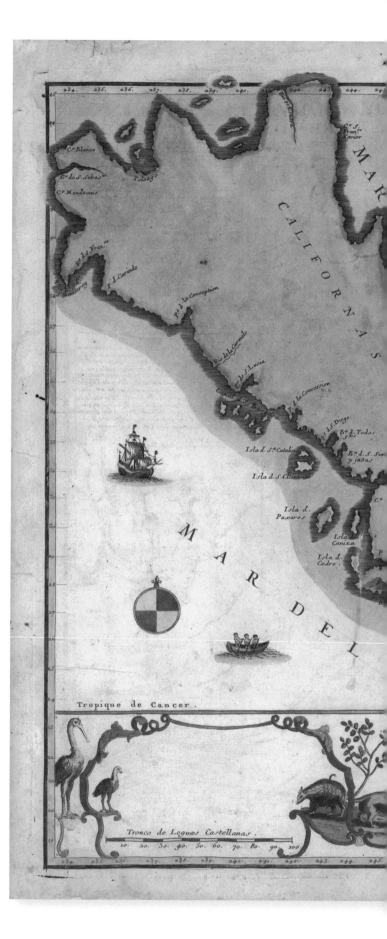

LA CALIFORNIE ou NOUVELLE CAROLINE.
TEATRO DE LOS TRABAJOS, APOSTOLICOS DE LA COMPA. E JESUS EN LA AMERICA, SEPTe.

Dressée sur celle que le Viceroy de la Nouvelle Espagne envoya il y a peu d'années a Mrs de l'Academie des Sciences. Par N. de Fer Geographe de sa Majesté Catolique

DON FERNAND CORTEL, Decouvrit le premier la Californie en 1533. et decendit en vn endroit qu'il nomma le Port de N. Dame. En 1535. le General François de Alarcon, passa dans la Californie avec quelques grands Vaisseaux a ses depens en 1586. et y passa avec cinq Religieux de St François. Le même y retourna en 1602. aux depens de Philipe III. Roy d'Espagne, avec trois Religieux Carmes, et debarqua a la Côte oposée de Californie en 1608. ou il receu vn ordre du Roy son Maire de passer sur la Côte Occidentale de cette Isle, et d'etablir vne Colonie au de Monteray, situé dans la Partie Septentrionale. Le Capitaine jean Eyturbie passa deux fois dans la Californie en 1615. Le Capitaine Francois Ortega y passa trois fois a ses depens; la premiere en 1632. la deux. en 1633. et y mena deux Prêtres, et la troisieme en 1634. Le Capitaine Carbonelly passa quelque apres le Capitaine Ortega. Le Gouverneur de Cinalea dans la Nouvelle Espagne appellé Cestin de Cana alla a la Californie en 1642. avec le P. Iacinte Cortes. de la Compag. de Iesus. L'Amiral Pierre Porter Casanat y Fut l'an 1644. et debarqua a la Baye de St Barnabe, et en 1648. et 49. il passa vne Seconde fois avec quelques PP. de la Compagnie de Iesus, et avanca de quelque degres plus au Nord. L'Amiral Don Bernard de Bernal de Pinadero y aborda en 1664. avec deux Navires du Roy d'Espagne, et en 1672. il receut vn Ordre d'passer vne Seconde fois. Le Capitaine François Lucenilla y alla avec deux Vaisseaux en 1668. equipé a ses depens, avec deux Religieux de St François. L'Amiral Isidra de Atondo et Antillon fit construire trois Vaisseaux a Cinaloa qui couterent plus d'vn demi million a Charles Second Roy d'Espagne avec les quels il passa a la Californie, et aux Côtes oposees en 1681. 82. 83. 84. et 1685. il avoit avec luy trois PP. de la Compagnie de Iesus, avec les quels il entra dans le Pays et en prit possession au nom et pour sa Majesté Catoliq. aux quels il donna le Nom de Nouvelles Caroline, en faveur du Roy son Maitre; Ce fut dans cette mission qu'on Instruisit dans les Principaux misteres de nôtre Religion quelques nations de cette Isle qui demandoient le Batême. Et dans ce tems le Capitaine François Ytamara passa a la Californie, et les naturels du Pais luy demanderent avec empressement des PP. de la Compagnie de Iesus en 1694. En 1693. 1e 19. Decembre quelques Vaisseaux partirent de la Province de la Pimerie dans le nouveau Mexique aux depens de la Nouvelle Espagne, Ils découvrent les lieux de Califor. nie qu'étoient les plus voisins de cette Province de Pimerie, et le 27. de Novembre 1694. a la hauteur de 34. degres, on découvre l'Agreable et Profonde Riviere du Coral qui prend sa Source dans la Pimerie, et qui se vient jetter dans le Bras de Mer de la Californie. Cecy a eté fait en 1695. pour le Viceroy de la Nouvelle Espagne, afin de poursuivre le dessein de la Conqueste et de la Conversion des Isles Californes ou Nouvelles Carolines.

A PARIS dans l'Isle du Palais a la Sphere Royale

GRAN TEGUAIO

GRAN QUIVIRA

MOQUI

NUEVO

MEXICO

APACHERIA

PIMERIA

SUMAS

JANOS

Elpasso

CALIFORNAS Ó CAROLINAS

CAROLINAS Ó

PARTE DE LA CALIFORNIE

Culiacan

Cinaloa

MAYO

Zacatocas

Tlacotes

Guadalaxara

Compostela

PARTE DE LA NUEVA ESPANA

GOLFE

DE

MEXIQUE

S. Luis Potosi

Aguas Calientes

S. Luis d'la Paz

St Felipe

MEXICO

Villa Rica

Queretaro

Valladolid

Puebla

Lac Chapala

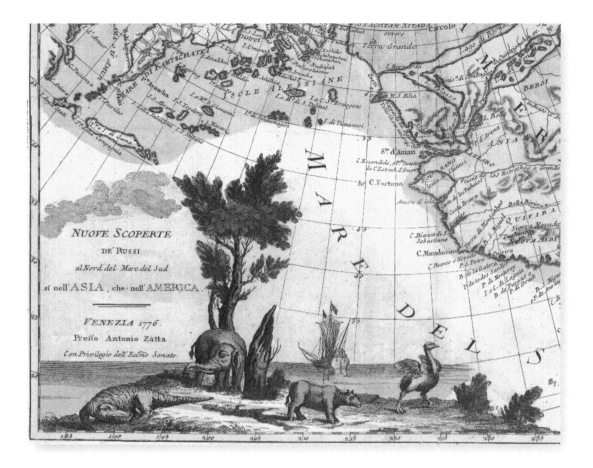

NUOVE SCOPERTE DE' RUSSI AL NORD . . . , VIGNETTE.

In the lower left corner of this 1776 map of the northern Pacific Basin, below the title, the Venetian map-maker and publisher Antonio Zatta (active 1750–1804, d. 1804) prominently depicts an alligator, an elephant, a rhinoceros, and an ostrich. These beasts from warmer climes are misplaced on a map that focuses on far northeastern Siberia and western North America. Here again a distant entre-preneur has exoticized a map to enhance its appeal and salability.

MAP OF THE BRITISH EMPIRE IN NORTH AMERICA . . .

Equally exotic imagery is prominent on the large, twenty-sheet 1733 map of the *Map of the British Empire in North America . . .* (*opposite*) by the geographer, mathematician, author, and publisher Henry Popple (d. 1743). The Americas are allegorically represented in a vignette surrounding the cartouche (*enlarged overleaf*) by three half-naked "noble savages," a common subject on maps of this period, amid the regional flora and fauna. In the background, European merchants trade with Indians for the riches of the New World, and at the right extreme of the vignette are tobacco plants, chests, and rum barrels. Popple had become intimately aware of the commerce with the European American colonies as a clerk at the British Board of Trade and one of its agents in the West Indies from 1727 to 1739. A relatively passive alligator acts as a footrest for the central female native figure, who rests one foot on a human head with an arrow in it, perhaps shot by the warrior who bears a bow at the lower left. This scene may symbolize Native American resistance against the Europeans. Similarly, the right-most female figure, who is caressing a child, gestures apprehensively toward the Old World traders. The Americas also are represented by tropical palms and northern fir trees. But note that, anomalously, a parrot perches in a fir tree. Two monkeys play near the alligator, and a shadowy wolf lurks behind the Indian archer.

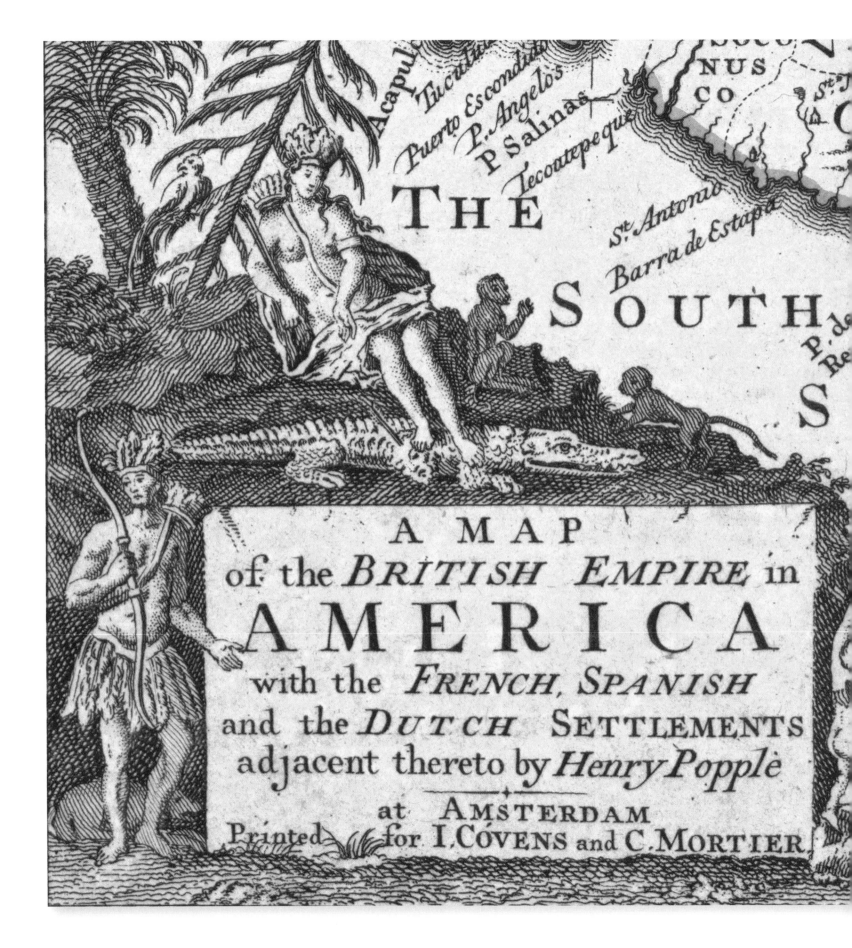

Acapul... Tucutu... Puerto Escondido5 P. Angelos P. Salinas Tecoutepeque

NUS CO

THE

St. Antonio
Barra de Estapa

SOUTH

P. de Re...

S

A MAP
of the *BRITISH EMPIRE* in
AMERICA
with the *FRENCH, SPANISH*
and the *DUTCH* SETTLEMENTS
adjacent thereto by *Henry Popple*

at AMSTERDAM
Printed for I. COVENS and C. MORTIER.

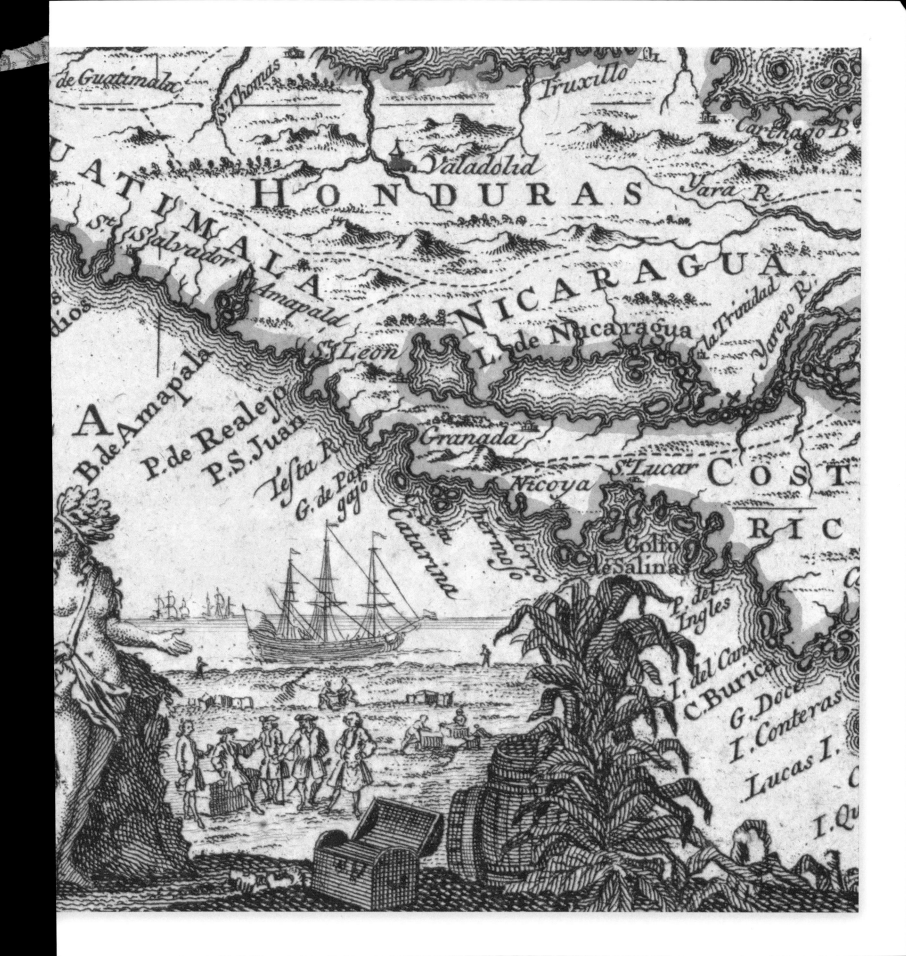

de Guatimala

S.Thomas

Truxillo

Carthago B.

Valadolid

HONDURAS

Yara R.

GUATIMALA

St.Salvador

S.Salvador

Amapala

NICARAGUA

la Trinidad

Yarepo R.

dios

St.Leon

L. de Nicaragua

A

B.de Amapala.

P.de Realejo

P.S.Juan

Testa R.

G. de Papa gayo

Granada

Nicoya

St.Lucar

COST

RIC

Santa Catarina

Golfo de Salina

I. del Ingles

I. del Can

C. Burica

G. Doce

I. Conteras

Lucas I.

I. Qu

NOVA SCOTIA AND CAPE BRETON

NEWFOUNDLAND

SCALE

10 20 40 60   100 Miles

LABRADOR

MINGAN DISTRICT

CANADIAN CHANNEL

ANTICOSTI ISLD

RIVER ST LAWRENCE

GULF OF

CANADA EAST

CHALEUR BAY

NEW

BRUNSWICK

ST. LAWRENCE

MAGDALEN IS.

PRINCE EDWARD ID.

BRETON

ISLAND

NOVA SCOTIA

BAY OF FUNDY

NEWFOUND

BAY OF ISLANDS

BAY OF ST GEORGE

FORTUNE

Miquelon I.

Langley I.

Sable I.

A T L A N T I C

Longitude 64 West from Greenwich

The Illustrations by A. Fussell & Engraved by J. Rogers

JOHN TALLIS & COMPANY, LONDON & NEW YORK.

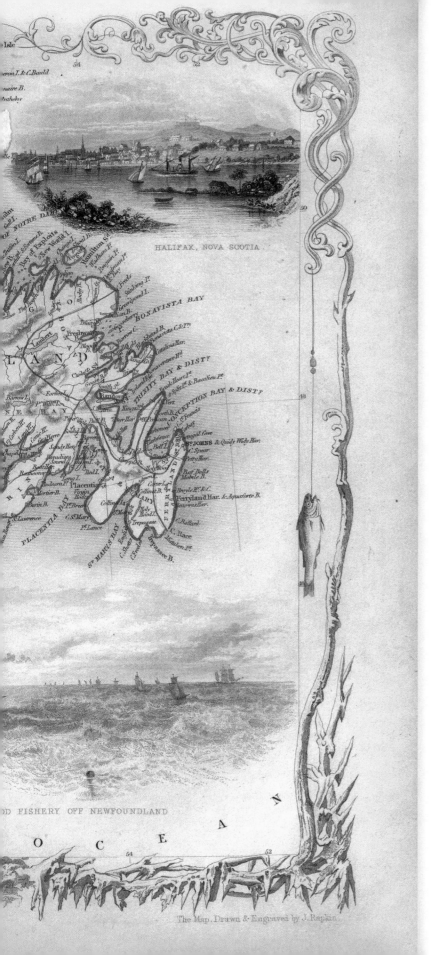

HALIFAX, NOVA SCOTIA.

COD FISHERY OFF NEWFOUNDLAND

The Map, Drawn & Engraved by J. Rapkin.

NOVA SCOTIA AND NEWFOUNDLAND.
Decorative animals persisted, especially
on maps of the Americas, throughout a
large part of the nineteenth century. But
the variety of native fauna and the way it
was portrayed reflected change over time.
For example, the 1851 map *Nova Scotia
and Newfoundland,* printed by London map
publisher John Tallis, shows two domestic
dogs of the types helpful in the exploration
and settlement of northern Canada and
Alaska; one is definitely a Saint Bernard.
These wonderfully unique canine characters
bring to mind the heroic dogs featured in
American novelist Jack London's northern
adventure tales, published half a century
later, including *The Call of the Wild* (1903)
and *White Fang* (1905). The various Australia
maps from the same Tallis atlas show many
of that continent's curious creatures—for
example, a kangaroo on the *Victoria or
Port Phillip* map and an elusive Tasmanian
devil on the *Van Diemen's Island or Tasmania*
map. Tallis's animals were intended to go
beyond merely being decorative to also
playing an educative role.

HOLT'S NEW MAP OF WYOMING . . . Some maps depict common farm animals, such as the almost flawless horse and several cattle prominently exhibited in the upper left corner of *Holt's New Map of Wyoming . . .* (1883), "compiled by permission from official records in U.S. Land Office" by George L. Holt and by Frank and Fred Bond, "draftsmen." Holt was a bookseller and stationer in Cheyenne, Wyoming; little is known of the Bonds. At the time of the publication of this map, the era of the American frontier was coming to an end, and the Wyoming Territory would soon become a state (it became the forty-fourth state of the United States in 1890). The domesticated animals and the counterbalancing factory scene in the upper right corner of the various editions of this map were meant to help convince officials in Congress and the White House, and those planning to move to Wyoming, that the territory was ready for statehood.

MONTANA IN 1864.

BIRD'S EYE VIEW OF THE CITY OF MONTANA, BOONE CO., IOWA 1868. This detail of a perspective bird's-eye map of Montana, Iowa (present-day Boone County, Iowa), by prominent American panoramic artist Albert Ruger (1829–99), also shows pastoral scenes of farming in America's heartland.

The Map as Animal

Animal life regularly populated—sometimes even overpopulated—maps during the golden age of cartography in the seventeenth century. And as in the case of the double-headed Hapsburg imperial eagle on Philipp Eckebrecht's 1630 map *Noua Orbis Terrarium Delineatio . . .* (see pages 26–27), on occasion animals even seem to take control of maps. Sometimes they also became the maps. For instance, in the *Itinerarium Sacrae Scripturae* (1581)—a sort of illustrated travelogue of Old and New Testament religious figures by the German theologian Heinrich Bünting (1545–1606)—there is a map of Asia in the shape of Pegasus, the winged horse of Greek mythology (see *below*). Other maps show the world as a three-leaf clover and Europe as an empress. [28]

Like other noncartographic elements included on maps, bestiary figures can often serve more than one function. They are, of course, decorative and were employed to beautify and boost the popularity (and value) of maps. At the same time, newly discovered and sometimes fantastical creatures explained and educated map viewers about the far-flung regions depicted. Even more than plants or ships, animals—such as the Holy Roman Empire's eagle, the Netherlandic lion, and the North American beaver—could stand as commanding allegories of the lands they represented and reinforce the political-economic presence of those countries on maps.

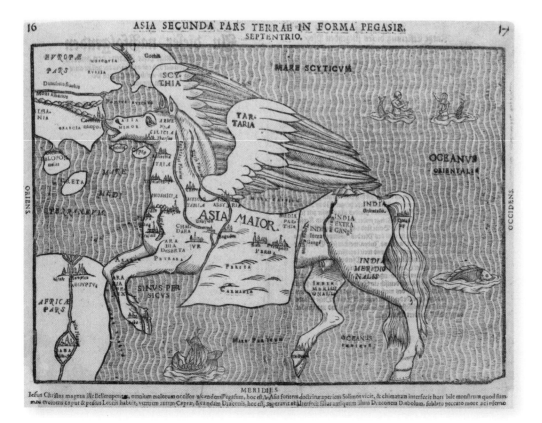

ASIA IN THE FORM OF PEGASUS. Heinrich Bünting, from *Itinerarium Sacrae Scripturae*, 1581.

LEO BELGICUS. By far the most elaborate and popular animal maps are those of the genre Leo Belgicus, depicting the Netherlands as a lion rampant. These maps derive from an original, shown here, by Baron Michael von Aitzing (c. 1530–98) and Frans Hogenberg (c. 1538–90) to illustrate Aitzing's history of the Low Countries, *De Leone Belgico* (Cologne, 1583). It was closely copied by, among others, Jan van Doetecum the Younger (active 1592–1630) in 1598 and Henricus Hondius in 1630. Versions of Leo Belgicus commonly appeared into the second decade of the eighteenth century.

LEO BEL-GICUS

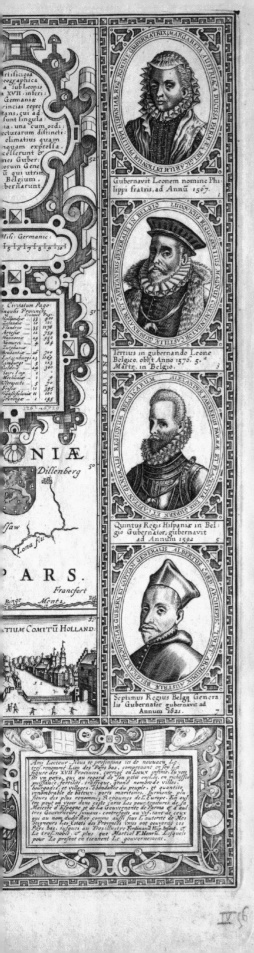

LEO BELGICUS. Nicolaes Visscher (1587–1652), the Flemish founder of a family publishing enterprise that lasted into the eighteenth century, issued his finely executed *Leo Belgicus* in Amsterdam in 1650. A Leo Belgicus map—with its transformation of the landmass into a lion and the addition of other decorative components—was meant to be an aesthetically pleasing objet d'art, a single-sheet artwork to be purchased by the frugal seventeenth-century patriotic burgher and proudly hung on the wall of the family home. But is the primary function of a Leo Beligcus map cartographic or ornamental? Is it a map or a cartifact (cartographic artifact)—something with cartographic aspects, such as the necktie decorated with compass roses cited in chapter 1, but whose principal function is not that of a map?

Visscher's *Leo Belgicus* is clearly a product of the time when Dutch economic power was peaking. The map exudes national pride and an awareness of history. The lion is indeed rampant, with its back to a sea filled with Dutch ships and facing right toward most of the rest of Europe; the lion's curled tail seems to be protecting those ships. Portraits of provincial governors and House of Orange *stadtholders* surround the lion. The distinctions made between northern and southern governors and the insets of the court of Brussels and the court of Holland below the lion's regally raised paw make this the first map of the Netherlands to highlight the separation of the country into northern (Dutch) and southern (Belgian) constituencies. Yet it was intended to and did appeal to the proto-nationalistic citizens of both parts as Netherlanders.[29]

The high-quality engraving, skillful application of color, and secondary symbolic embellishments also contribute significantly to the total effect. Although the map is not distorted, it is encompassed by the lion and thereby becomes a central part of the larger, quite attractive composition. The coming together of artistry, geography, and political history in Visscher's *Leo Belgicus* seem to suggest that its value was more as a pictorial cartifact. Artistic maps of the Leo Belgicus type were more readily affordable than the large, colored, multisheet Dutch wall maps of the day, which could cost as much as an oil painting. At approximately one-twentieth the price of those larger maps, examples of the Leo Belgicus sold for what was considered "still a tidy sum."[30]

Natives of British Asia and of East Indian Islands

PISCO:
ATVS
S.

6

The Peoples of the World

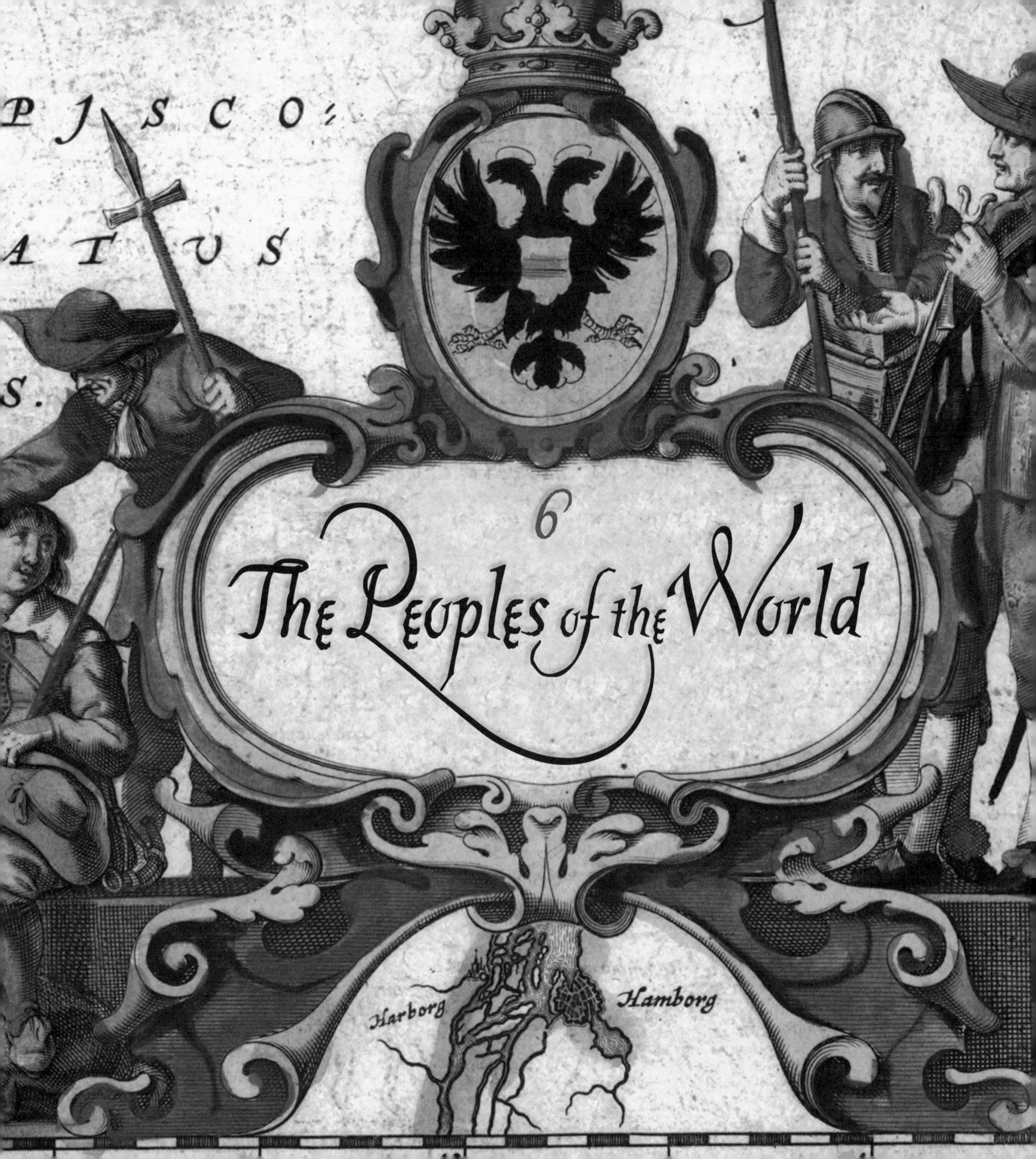

Harborg   Hamborg

"For the child, in love with maps and prints. . . .
Ah! How wide is the world by lamplight!"
—Charles Baudelaire, "The Voyage" from
The Flowers of Evil, 1857 [31]

hile the cartographic flora and fauna of distant lands and seas readily widened the world of map users—adults and children alike—it was humanity's glorious diversity as represented on maps that surely enchanted them most. Before the momentous "Columbian age" of European discovery and exploration in the fifteenth and early sixteenth centuries, the African, Asian, and European peoples of the Old World were at least somewhat known to one another. Political, social, economic, military, and biological interactions between them had been recorded for more than two millennia. Although a certain sense of "otherness" yet existed between them and they were still very curious about one another, by the time of the first Iberian voyages beyond Europe, these peoples were all part of the then-acknowledged human community, of what the British geopolitician Sir Halford Mackinder (1861–1947) later called the "World Island" of Africa, Asia, and Europe.

But after the first contact of 1492, the previously unknown peoples of the New World created quite a stir in the Old, especially in Christian Europe. At first, not only was the existence of these mysterious foreigners questioned; the reason for their existence came under scrutiny as well. In the late fifteenth century, the then-fundamentalist Roman Catholic and Orthodox churches believed that all the peoples of the earth were to be found in the divinely revealed word of the Bible. Were they not all descended from the three sons of Noah—Shem, Ham, and Japheth—who came to inherit what would be Africa, Asia, and Europe after the Old Testament flood?

The Bible, however, made no mention of the New World and its peoples, those who would collectively come to be called Indians by Columbus and his successors in exploration. So who were they? The progeny of a lost tribe? The outcast spawn of Satan and his followers?

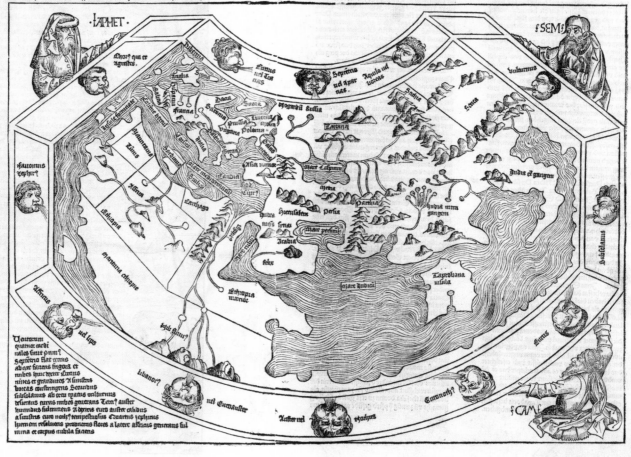

DAS ANDER ALTER DER WERLT. This world map is by German historian and cartographer Hartmann Schedel (1440–1514), author of the *Liber Chronicum*, better known as the *Nuremberg Chronicle,* a biblical world history published in 1493 in Nuremberg. The map here, which appeared in the German edition of the *Nuremberg Chronicle*, is illustrated with a range of bizarre beings believed to inhabit unexplored regions of the world. Such imagery reinforced the contemporary perception that peoples of distant lands were "other": mysterious and savage.

PREVIOUS PAGES: *Right*, **RVSSIAE . . . ,** from *Le Theatre du Monde, ou, Novvel Atlas.* Willem Janszoon Blaeu, Amsterdam, 1647.

Left, **BRITISH EMPIRE . . .** John Bartholomew, Edinburgh, c. 1850.

Viking Contact

The earliest documented Old World contact with the New was by the Norse in their movement across the North Atlantic at the end of the tenth century. Eventually, Vikings such as Leif ("the Lucky") Eriksson and Thorfinn Karlsefni sailed from their colonies in Greenland to the east across the Davis Strait and landed in present-day Labrador and Newfoundland in northeastern Canada—designating them as Markland and Vinland. In the process, they also first encountered indigenous peoples, whom they called *Skraelings*. Yet this contact created nowhere near the commotion that would occur five hundred years later. Scandinavia was only beginning to become Christianized in the tenth century, so most of the early Norse explorers were likely still pagan and not familiar with the Bible. Also, the Vikings had previously interacted with the Lapps in northern Scandinavia and, more important, the Inuits (Eskimos) in Greenland, so the *Skraelings* would not have seemed so strange to them. The accounts of the *Skraelings* were generally lost to the rest of Europe as they were passed down in the insular, mainly oral tradition of Norse sagas.[32]

Although the Vikings used the compass, they sailed according to mental maps and did not engage in formal cartography. As reflected on the cartography of the fifteenth and sixteenth centuries and beyond, it took more than a half century after Columbus for Christian Europe to incorporate the Indians into its now rapidly changing Western worldview and to decide the Indians' fate. Eventually, after much debate and the entreaties of such enlightened individuals as the Dominican friar Bartolomé de las Cases, a personal friend of the Columbus family, the Catholic Church issued a papal bull confirming that Indians were human, rational beings with souls. They were therefore ripe for conversion and could no longer be enslaved. The Native American population had also been decimated by diseases brought to them by the Europeans. Sadly, kidnapped Africans fared better in resisting European diseases; over the next four centuries they were enslaved in vast numbers to meet the growing demand for labor in the ever-expanding European colonies of the New World.

A Harper's Weekly *illustration from 1875 shows a Viking ship approaching Greenland.*

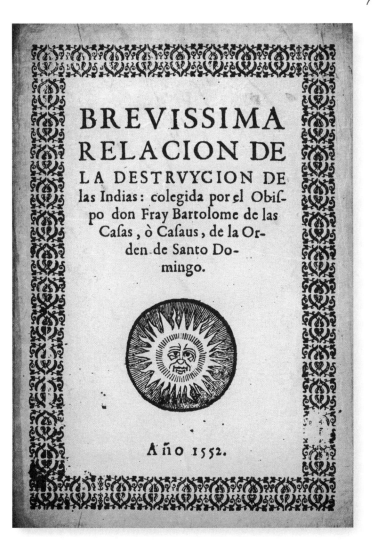

BREVISSIMA RELACION DE LA DESTRVYCION DE las Indias: colegida por el Obiſpo don Fray Bartolome de las Caſas, ò Caſaus, de la Orden de Santo Domingo.

Año 1552.

BREVÍSIMA RELACIÓN DE LA DESTRUCCIÓN DE LAS INDIAS, TITLE PAGE. The Spanish historian and Dominican friar Bartolomé de las Casas (c. 1484–1566) wrote this missive about the mistreatment of the indigenous peoples of the Americas in colonial times and sent it to Prince Philip II of Spain. The title translates as *A Short Account of the Destruction of the Indies*; it was written in 1542, published in 1552.

Native Peoples

By and large, on fifteenth- and sixteenth-century maps sub-Saharan Africans and Native Americans were portrayed quite differently from Asians, North Africans, and Europeans. Because of the acknowledged long history and achievements of the once high civilizations of Asia and the Middle East, the contemporary descendants of these cultures were rendered with due respect by European mapmakers, though the Europeans believed themselves to be superior. After all, countries such as Egypt and China had been "civilized" for a long time, whereas sub-Saharan Africans and Indians—whose cultural achievements were yet unknown—were erroneously thought to be "inferior," living in varying states of savagery and awaiting the blessings of progress borne by the Europeans. As mentioned previously, that the Native Americans were not mentioned in the Bible initially created significant consternation; the first contact with them was shocking. The notion that they were savages who could be easily defeated—they were depicted as such on maps—reinforced the ensuing European invasion of the New World. However, in part because the Europeans did not enslave Native Americans, the metaphorical depictions of Indians on maps soon evolved differently from those of Africans.

PROVINCE DE QUANG-TONG AND GVINEA PROPIA . . . , DETAILS. Chinese merchants and workers in the cartouche of d'Anville's map of China (1737) (*left*; see full map on page 101) are portrayed as somewhat exotic relative equals to their European customers, whereas the Africans in their village on the Homann Heirs map of West Africa (1743) (*below*; see full map on page 99) are shown as much more primitive and living closer to nature.

VALLARD ATLAS, CHART 1, TERRA JAVA, DETAIL. Asian figures are given a similar deference in the charts of the *Vallard Atlas* (1547). Those Asians rendered as least sophisticated are the residents of the East Indies on the chart of "Java" (*opposite*), yet they are not savages. They are portrayed fully clothed and as living rather idyllically, surrounded by lush vegetation that includes some of the spice plants that were of great value to the Europeans. The lifestyle of the Javanese was viewed as a product of their location, the subtropical climate, and such pursuits as farming and hunting.

Straita
thoengros
baffalonga
ropelyo
Riogrand
S: vicolas
Is: S fransois
cap Vetoga
bancodent
Illagrossa
Rio fondo
tres illes
capdenble
Riotinior
costoangerogo
Riobassa
laffagramp
scila

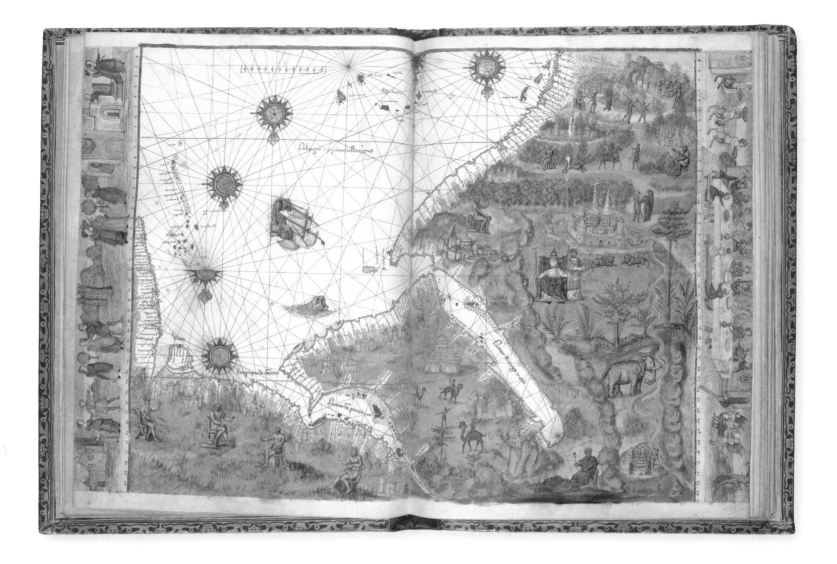

VALLARD ATLAS, CHART 4, ARABIAN SEA, RED SEA, AND PERSIAN GULF. The followers of Islam are seen as being of an even higher level of civilization; the closer to Europe, the more civilized in manners and dress they become, as illustrated on the chart of the Arabian Sea area (shown here), the Adriatic Sea area (chart 14), and the Aegean Sea region (chart 15). They were at times the trading partners, rivals, and/or sworn enemies of the Europeans.

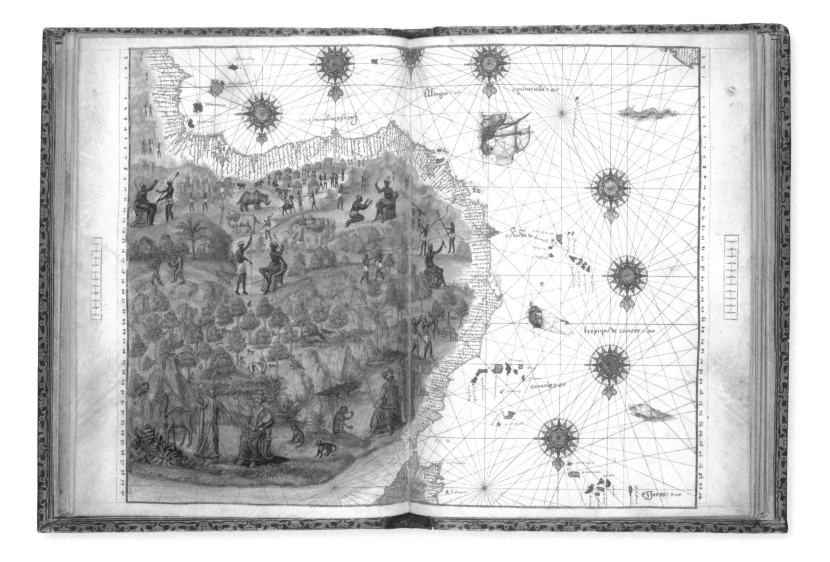

VALLARD ATLAS, CHART 7, NORTHWEST AFRICA. Perhaps nowhere in the *Vallard Atlas* is the contrast between the more urbane and the less civilized as clearly apparent as on the chart of northwest Africa. In the north (the bottom left of this south-oriented chart), along the Mediterranean Sea coast, three elegantly robed Arabs appear. To the south, separated from the Muslims by the Atlas Mountains and forest (but, oddly, no desert) and along the northern "slave coast" of the Gulf of Guinea, Africans wear loincloths, an ancient mode of dress in warm climates that sixteenth-century Western Europeans thought to be indecent. There are those who hunt slaves for the Portuguese and Arabs, such as the residents of Timbuktu and the Kingdom of Benin, whose towns are drawn to look like European ones, and those who would be the slaves, shown amid native villages and African beasts. The Africans on the chart of southern Africa (chart 5) are drawn in a comparable primitive fashion.

TARTARIAE SIVE MAGNI CHAMI REGNI TŸPUS. In the upper left corner of the 1570 map of Asian Tartary (western China and Mongolia) from Abraham Ortelius's *Theatrum Orbis Terrarum* there is a vignette of a Mongol warrior, perhaps the illustrious Genghis Khan, in front of his distinctive tent, a yurt. The drawing is both a decoration and a reminder of the great Mongol conquest that began with Khan in the late twelfth century. By the time of the death of his grandson Kublai Khan in 1294, when the Mongol Empire was at its height, fearsome Mongol fighters—streaming out of their Central Asian homeland along the Gobi Desert—had conquered more than 80 percent of Eurasia, including China and Persia, along with large portions of India and Russia. Three centuries later, the Mongols were still remembered with respect and trepidation. Similar yurts fill in the open space across the northwest of Ortelius's 1584 China map based on the work of the Portuguese Jesuit missionary and mapmaker Luís Jorge (Ludovico Georgio) de Barbuda (active 1575–1599).

CIVITATES ORBIS TERRARUM, BYZANTIUM, NUNC CONSTANTINOPOLIS (TOP), CAIROS (BOTTOM). Inspired and buoyed by the success of Ortelius's atlas, in 1572 the Germans Georg Braun (1541–1622) and Frans Hogenberg (c. 1538–90) began publishing their *Civitates Orbis Terrarum*, an atlas of bird's-eye views of the great cities of the world that was expanded until 1617. Braun, a topographer, edited the atlas and added many details to the maps, most of which were then finely engraved by Hogenberg. Across the bottom of the map of Constantinople (1572), portraits of Turkish sultans are magnificently arrayed, separated by a depiction of a sultan on horseback with some of his entourage. At the bottom lower left of the Cairo map (1572), Muslims in perceived local costume go about their daily pursuits, including picking dates.

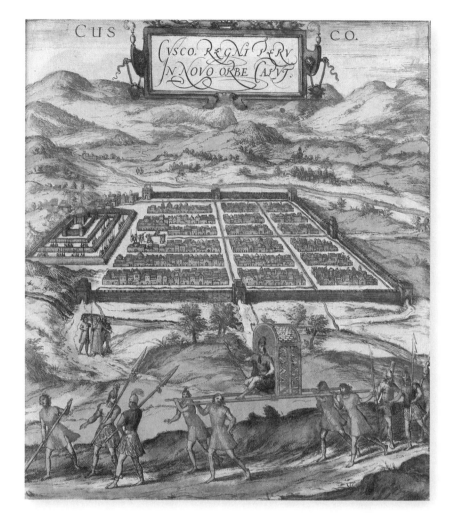

CIVITATES ORBIS TERRARUM, MEXICO AND CVSCO. Along the bottom of Braun and Hogenburg's combined one-sheet maps of Mexico City and Cusco (Cuzco, Peru) (1572)—shown here separately—outside the walls of European-style cities, citizens of these higher American Indian civilizations are represented in supposed native dress. On the Cusco city plan (*above right*), an Inca king, perhaps Atahualpa, who was betrayed and executed by the Spanish conquistador Francisco Pizarro in 1533, is carried on the shoulders of his subjects in a vehicle reminiscent of a sedan chair. Throughout the atlas, attractive additions such as these helped convey to viewers something of the energies of the major cities mapped.

AFRICAE NOVA DESCRIPTIO. On Willem Blaeu's beautiful, influential, and much-copied 1635 edition of his map of Africa, *Africae Nova Descriptio* (see full map on pages 78–79), a caravanner leads a camel near the West African trade center of Timbuktu (in present-day Mali) (detail enlarged *below*); beyond this depiction only animals adorn the continent and fill in its blank spaces. However, the map's left and right margins are composed of portraits of couples in local dress from across Africa (shown here together, *right*). These pictures reflect not only the diversity of the peoples of Africa but also the range of their levels of progress in terms of Western mores.

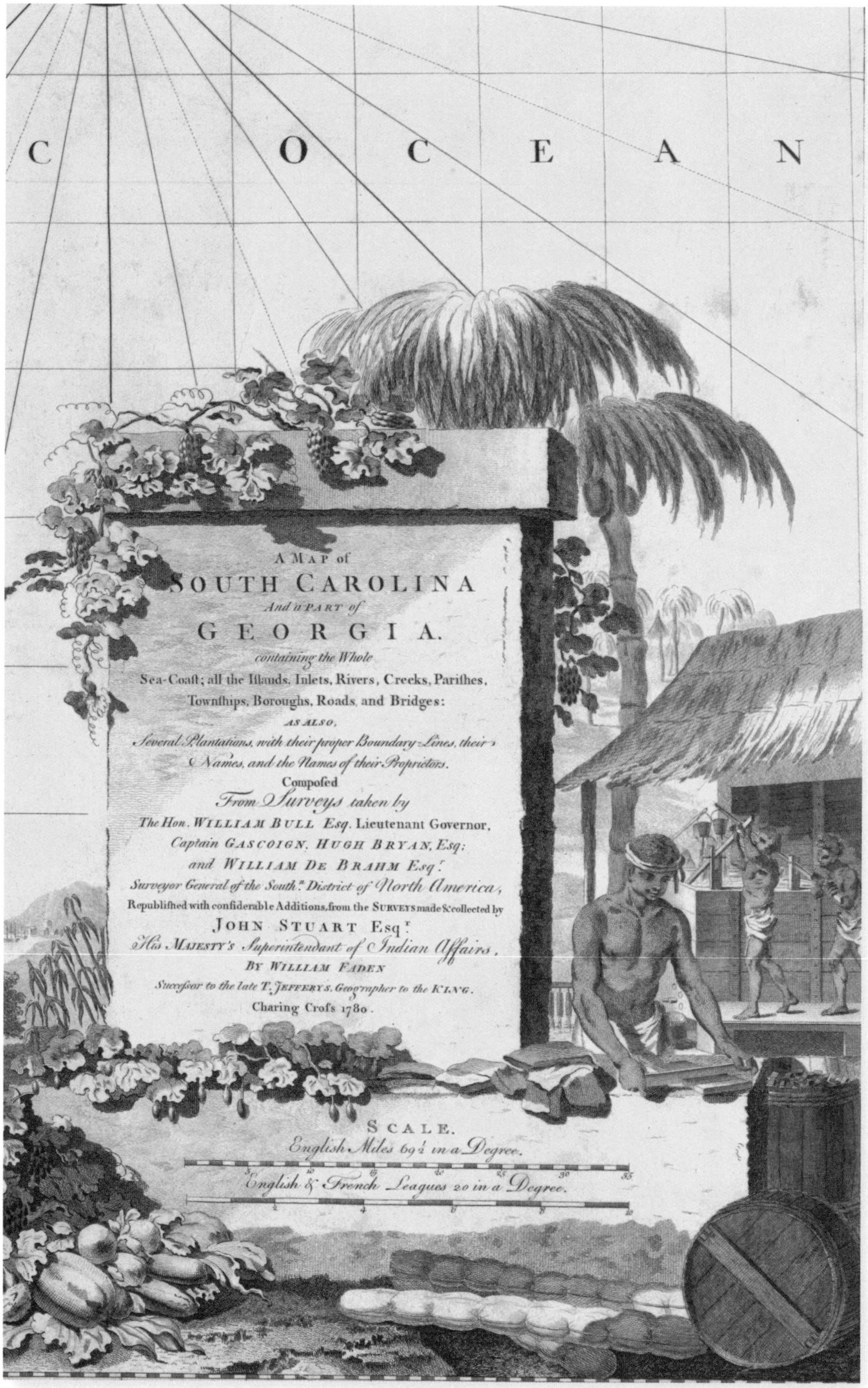

A MAP OF SOUTH CAROLINA AND PART OF GEORGIA, CARTOUCHE. Images of Africans enslaved, especially in the New World, began to appear on maps near the end of the seventeenth century. Although William Faden (1750–1836), a leading British cartographer and geographer to King George III, is best known for his excellent maps of the American Revolution, his 1780 two-sheet map of South Carolina and Georgia depicts an incongruous slavery vignette as part of the title cartouche in the lower right corner, shown here. Depicted are three strong, healthy slaves cheerfully engaged in various labors, including what appears to be tanning hides.

Cannibals and Other Monstrous Beings

Constituents of mythological "monstrous races" materialize on numerous examples of Renaissance cartography. Renaissance mapmakers were inspired by references to mythical hybrid peoples from classical Greek and Roman geographies and medieval art, literature, and cartography;[33] those references in turn were based on alleged sightings and accounts gathered by early travelers from Alexander the Great to Marco Polo. Early catalogs of these races were to be found in the first-century *Natural History* of Pliny the Elder and the third-century *Collectanea rerum memorabilium* ("Gallery of Wonderful Things") by Julius Solinus. Since by the late sixteenth century such creatures were still part of the body of popular belief but had not actually been discovered anywhere in the rapidly expanding known world, they were increasingly relegated to the yet-to-be-explored mysterious hearts of Asia and, especially, Africa by such mapmakers as Waldseemüller (e.g., *Carta Marina,* 1516) and those of the *Vallard Atlas.*

VALLARD ATLAS, CHART 7, NORTHWEST AFRICA, DETAIL. In the far left center of the Vallard chart of northwest Africa (see full map on page 139), below one of the representations of a seated native ruler with a subject, two members of ancient monstrous races are shown. One is of the dog-headed *cynocephali*, and the other is of the *blemmyae*, headless beings with faces in their chests.

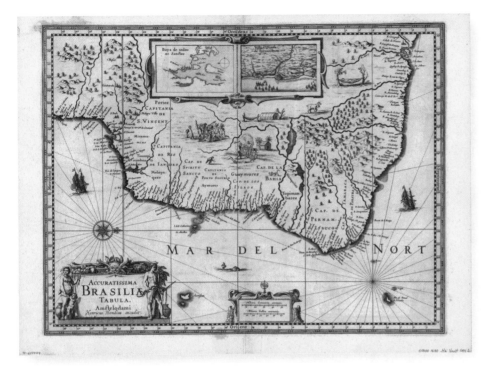

ACCURATISSIMA BRASILIAE TABULA. Another monstrous race, the *anthropophagi*, was reported to be cannibalistic. True and untrue accounts of native cannibalism came into Europe from across much of the Americas in the sixteenth and seventeenth centuries, and cannibals occasionally appeared on the New World cartography of the time. On the map of Brazil, *Accuratissima Brasiliae Tabula* (Amsterdam, 1630), by Dutch cartographer Hendrik Hondius (1597–1651), there are several such interesting pictures of South American Indian life in the still largely unknown interior (see *below*). In addition to two native dwellings, Hondius provides a scene of tribal warfare, with the naked victors making prizes of their equally naked enemies' heads and bodies. Farther to the right, a group from the conquering tribe is butchering the body of a captive and, in preparation for eating the parts, roasting them on a grill. The drawing hints at something of the ritual nature of cannibalism, if not its periodic necessity because of hunger.

DIE NEUWEN INSELN . . . In Sebastian Münster's woodcut New World map (1540 to 1628) from his *Cosmographia* (see full map on page 43), a leg and a head gruesomely hang from a tree in eastern South America, with the notation "Canibali" above.

TABULA TERRA NOVA, INSET DETAIL. The relatively famous (or infamous) inset from the 1522 edition of Martin Waldseemüller's *Tabula Terra Nova* (see full map on pages 30–31) is a good example of the type of pictures of Native Americans that appeared in Europe during this early period of contact. Placed in the white space of the *terra incognita* of South America, it depicts the rumored and reported cannibalism of the New Worlders; they are dismembering a female victim and making a meal of her and apparently of others' arms and legs. The figures strangely, if understandably—Waldseemüller had not actually seen any cannibals, nor had many other Old Worlders—look like ancient, semi-naked northern Europeans. This scene used nakedness, the European fear of being eaten (especially alive), and the taboo against cannibalism to underscore the savagery of the Indians.[34]

VALLARD ATLAS, CHART 9, NORTH AMERICA, EAST COAST. On the New World charts (9 through 12) of the *Vallard Atlas,* the Native Americans are engaged in such pursuits as hunting, farming, and mining (possibly for the Europeans), trading with each other and the Europeans, and even warring among themselves, but no cannibalism is depicted. They are often occupied in these pursuits in the presence of explorers from France (e.g., Jacques Cartier may be on the map shown here), Spain, and Portugal, some of whom are observing the natives. As on numerous maps of this period, the natives in the *Vallard Atlas* have European features but are easily distinguishable because of their full or partial nudity and their behavior. Not only are the Europeans fully clothed in the fashionable dress of the day, but they are often shown bearing swords and firearms. With their dress, technology, and look of superiority, the Europeans clearly overshadow the natives.

VIRGINIA . . . In 1612 a more stylized (and even generic) illustration of an Indian appeared on John Smith's map of Virginia (see full map on pages 6–7). The posed figure is based on an image of an Algonquin-speaking Secotan from the Carolinas (once part of the colony of Virginia) created by the prolific Belgian-German engraver Theodore de Bry. He engraved the image for *A Briefe and True Report of the New Found Land of Virginia* (1588), an account by English astronomer and mathematician Thomas Harriot (1560–1621) chronicling his expedition to the ill-fated Roanoke Colony, the English settlement in Virginia Colony whose inhabitants mysteriously vanished sometime between 1587 and 1590. [35] On Smith's map, the Secotan figure migrated north and was transformed into a Susquehannock.

A WEROANS, OR CHIEFTAIN, OF VIRGINIA. The figure on the left in this engraving by Theodore de Bry, from *A Briefe and True Report of the New Found Land of Virginia,* 1588, was the inspiration for the Susquehannock hunter (*above*) on John Smith's 1612 map of Virginia.

CODFISH MAP. A century later, on a c. 1717 map of North America (the Codfish Map) by Herman Moll (see full map and vignette on pages 60 and 61), Native Americans in both summer and winter dress are portrayed more realistically and as less wild-looking.

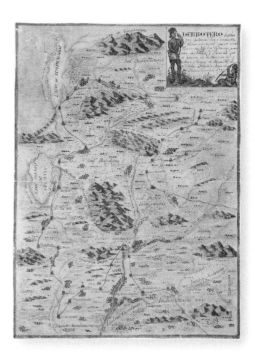

DERROTERO HECHO . . . An Indian in a similarly more realistic pose can be found in the title cartouche of the important hand-drawn manuscript *derrotero* (route map) of the Vélez de Escalante expedition into northern New Spain (led by Franciscan priests Atanasio Domínguez and Silvestre Vélez de Escalante) in 1777, more than a century and a half after the Smith map.

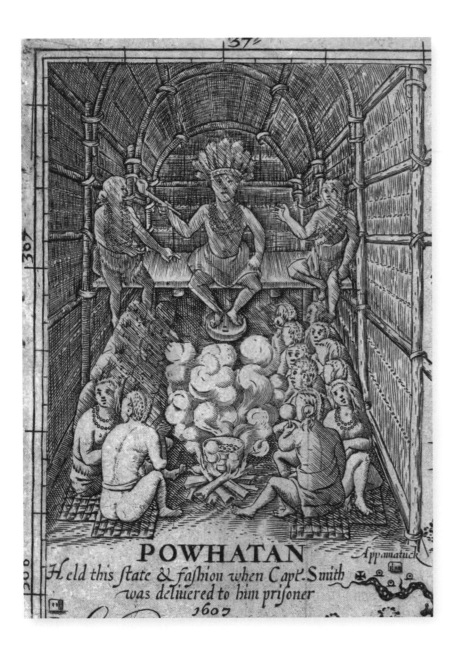

POWHATAN
Held this state & fashion when Capt. Smith
was deliuered to him prisoner
1607

Appamatuck

VIRGINIA . . . For want of a better understanding of Indian government, in the upper left corner of the Smith map, Chief Powhatan, leader of the Powatan tribe, is pictured sitting at the head of his council in a native longhouse, in a setting similar to the one in which an English monarch might hold court at the House of Lords.[36]

NOVI BELGII NOVAEQUE ANGLIAE . . . On the 1685 Nicolaes Visscher/Peter Schenk map of the mid-Atlantic region (see full map on page 16), two Mohican villages are arranged in an ordered, symmetrical European manner.

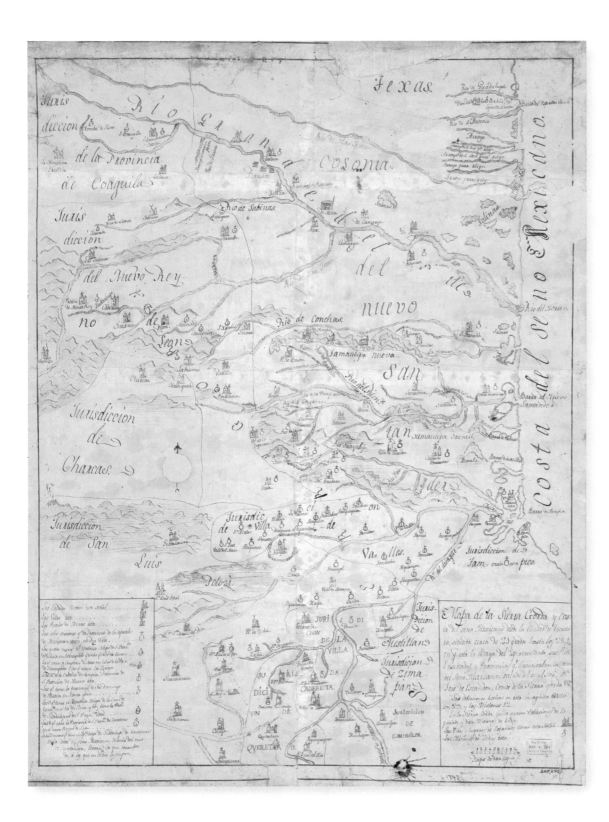

MAPA DE LA SIERRA . . .
The Indian pueblos of northern New Spain are still represented as graceful European Spanish towns on the c. 1747 map entitled *Mapa de la Sierra Gorda y Costa del Seno Mexicano . . .*, created for the governor of the New Spain colony of Nuevo Santander, José de Escandón (1700–1770). That Native Americans were in fact members of the human race took the best part of the sixteenth century to enter the European mind-set; the true facts about their appearance and culture remained largely unknown until at least the early eighteenth century, especially in the case of native peoples in the North American West and the South American interior.

The Noble Savage

It was out of a greater understanding (and misunderstanding) of the true nature of indigenous peoples that the idea of the "noble savage" emerged. In Western thought, the idea that the "noble savage" lived almost idyllically in the "state of nature" was a product of the Enlightenment of the seventeenth and eighteenth centuries, especially in countries that had ongoing imperial or national frontiers in the New World. This prevalent archetype was exemplified in the artistic and philosophical discourse of the time, including the literature and imagery—including maps—sparked by discovery and exploration of the New World, and such iconography prevailed into the first part of the nineteenth century and the era of Romanticism. British Enlightenment poet Alexander Pope delightfully captured the concept as it applied to Native Americans in the following lines from his work *An Essay on Man* (1732–34):

"Lo, the poor Indian! Whose untutor'd mind
Sees God in clouds, or hears him in the wind;
His soul proud Science never taught to stray
Far as the solar walk or milky way;
Yet simple nature to his hope has giv'n
Behind the cloud-topp'd hill, an humbler heav'n." [37]

The notion of the noble savage in the state of nature was influenced by the strangeness and "differentness" of the New World, but it also rested upon the foundation of noble "savages" from the Old World's past, such as the Celts, the Saxons, and the Gauls. The rather idyllic notion of the archaic European noble savage was used by would-be reformers of contemporary European civilization as a context for the plight of the lower classes, who faced poverty and inhumane living conditions. Modernization and development were seen as double-edged swords, deficient processes. Thus, almost from 1492 onward, Indians were frequently mentioned in assessments of the contemporary European social order. As historian Robert F. Berkhofer suggested, "The Noble Savage really pointed to the possibility of progress by civilized man if left free and untrammeled by outworn institutions."[38] On the other hand, David Weber elaborates that the state of nature was not necessarily a better place than civilized Europe and that in that state "nobility" did not necessarily connote "superiority"; actual Indian policy, for example, was always made more pragmatically. Nor was seeing the Indians as noble savages universal among Europeans; the French and Russians were most captivated by them and the Iberians less so.[39]

The "Noble Savage" is idealized in this cartouche from A New Map of North America, with the West India Islands, 1783, *published by Thomas Pownall, a member of the British Parliament who was the governor of Massachusetts Bay Colony from 1757 to 1760.*

WHICH COMPOSE
THE BRITISH DOMINIONS,
LAID DOWN *according to* THE LATEST SURVEYS,
and Corrected *from* THE ORIGINAL MATERIALS,
OF GOVER: POWNALL, MEM.ᴿ OF PARLIA.ᴹᵀ
1783.

BERMUDAS
or
SUMMER'S Iˢ

A NEW MAP OF THE COUNTRY OF LOUISIANA . . .

Henri Joutel (c. 1640–1735) was an officer who accompanied French explorer Robert de La Salle (1643–87) to the Mississippi River and Texas in 1684–87 and made his way back to his home in Rouen, France, via Quebec after La Salle was murdered by one of the mutinous members of the expedition. (The killing of La Salle is shown in the center of Nicholas de Fer's 1701 map of the Mississippi River country.) Joutel kept a meticulous manuscript journal of the La Salle expedition, which was used by explorer Pierre Le Moyne d'Iberville (1661–1706) during several trips he made to the

Mississippi and nearby Gulf Coast regions between 1699 and 1702. In 1713 Joutel published the account of his adventures, which included a map of Louisiana. In the lower left corner, two Indians hold up a buffalo hide that bears the title of the map. These muscular New Worlders, scantily clad in animal skins and with feathers in their hair, are at one with the state of nature; the one on the left bears a great club and the one on the right a crude bow. The Indian on the left even appears to have a tail, but perhaps it extends from the buffalo hide or his breechcloth. In their build, brutishness, and nakedness they are also reminiscent of the Europeans' archaic ancestors.

CARTE DU CANADA . . . In the cartouche of the 1703 Canada map by Guillaume Delisle (1675–1726) entitled *Carte du Canada ou de la Nouvelle France . . .* , an illustration of the Indian practice of scalping is placed diagonally opposite a scene of Native American baptism and Christianization. An intellectual child of the Enlightenment and a leading French cartographer of North America, Delisle was a member of the Royal Academy of Sciences and royal geographer; he was known for his stressing of accuracy.

Iroquois Allies

Herman Moll offered many images and comments on the primitive nobility of the American Indians. For instance, on the previously discussed 1715 Beaver Map of North America, this telling reference to the Iroquois (see commentary on page xxi) appears to the west of Pennsylvania:

"The Iroquois consist of four Cantons, Govern'd by so many Kings, and are all hearty friends to ye English: those Princes came into England in 1710 to offer their services agt ye French in Canada, and had it not been for ye miscarriage of our Expedition to Quebec in 1711 those people would have been of great service to us, for they joyn'd General Nicholson with 2,000 men on his March to attack Montreal."

Oddly, Moll indicates only "four Cantons" of the five-nation (eventually six-nation) Confederacy of the Iroquois, but he obviously has respect for them. Moll, a Whig, refers to the Iroquois' unique, quasi-democratic political system and values them as allies to the British (as they had been to the Dutch in the mid-seventeenth century) in Queen Anne's War (the North American theater of the War of the Spanish Succession, 1702–1713). He fails to mention that the French also had honored allies in the Huron Confederacy on the same frontier in successive wars.[40]

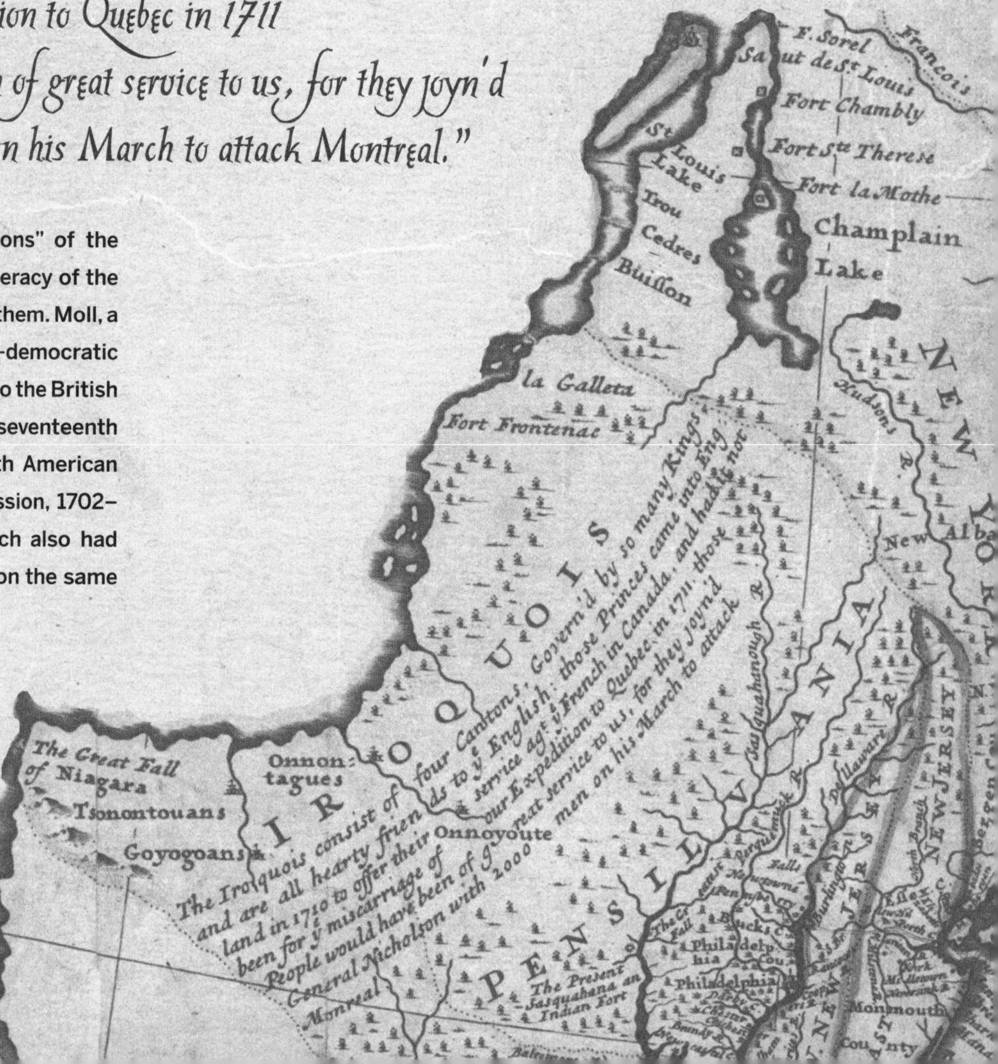

MAP OF THE UNITED STATES OF AMERICA. The "noble savage" of the Enlightenment preceded the "sentimental savage" of the Age of Romanticism. In the early nineteenth century, pictorial representations of Native Americans—whether on maps, on product labels (especially tobacco, see label *right*), or in graphic arts (such as play posters)—started to become overly romanticized and even sentimentalized. Consequently, in the upper left corner of the highly nationalistic *Map of the United States of America* (1850) by J. H. Colton (see full map on page 44), there is an idealized picture of now "Americanized" Indians fishing for salmon at Willamette Falls, Oregon, yet in a traditional way (*below*). This was a Manifest Destiny map that clearly showed the territorial gains made by the United States in Oregon and the Mexican-American War.

A tobacco label for Pocohontas Chewing Tobacco, 1867.

JOHNSON'S NEW ILLUSTRATED FAMILY ATLAS . . . An even more romanticized view of Native Americans is shown on the frontispiece to *Johnson's New Illustrated Family Atlas, with Descriptions, Geographical, Statistical, and Historical,* co-compiled by J. H. Colton and A. J. Johnson (1827–84) and published in 1862.

GUNN'S NEW MAP OF KANSAS AND THE GOLD MINES . . . Another map celebrating the "noble savage" is the magnificent *Gunn's New Map of Kansas and the Gold Mines . . .* (1859; the full map shown here is an 1867 edition). Otis B. Gunn was from Wyandotte County, Kansas, and his map and travel guide were published in response to the short-lived Kansas or "Pike's Peak" Gold Rush, which eventually moved westward, and to the hopes for Kansas statehood. Just inside the center of the left margin is an archaic rendering of a "noble savage," dressed in animal skins, carrying a bow, and perhaps still running free in the foothills of the Colorado Rockies on the western edge of the Great Plains (enlarged, *below left*). But almost at the center of the 1859 edition of the map, in the once-designated "permanent Indian country" of the 1830s—enclosed in a surveyed rectangular reservation labeled "Pottawattamie Reserve"— are several sentimentalized, "tamer" Indians (*below right*) admiring an allegorical bust of George Washington as father of his country (a standing figure of Washington as frontiersmen with his rifle contemplates the bust as well). Several other "reserves," such as that of the Sac and Fox tribes, are also clearly designated; the reserves often turned out to be temporary, as the Indians were constantly cheated out of their treaty lands. The ultimate message of Manifest Destiny is clear: the United States was expanding westward and the Indians must depart or "civilize" and assimilate.

Europeans and Historical Figures

Throughout the seventeenth-century golden age of mapmaking and beyond, depictions of Caucasian Westerners remained in striking contrast with those of Asians, black Africans, and Native Americans. Generally, Westerners were portrayed as distinct and superior. Many of the illustrations depicted specific individuals. The bearded figure in the black cape and red hat on the *Vallard Atlas* North America chart 9 (see map on page 149) may be French explorer Jacques Cartier (1491–1557).

UNIVERSALIS COSMOGRAPHIA SECUNDUM . . .

At the center of the top of Martin Waldseemüller's unique and significant twelve-sheet woodblock printed world map of 1507, *Universalis Cosmographia Secundum . . .* (see full map and further information on pages 32–35), two well-known, identifiable personages appear, one on either side of the two hemispheres of the globe. On the left side is a late-medieval representation of Claudius Ptolemy (*left*). The figure on the right is Amerigo Vespucci, Waldseemüller's "continental namesake" (see detail on page 35). Since Waldseemüller was well acquainted with Vespucci through reading and publishing his accounts, Vespucci's portrait should be regarded as more accurate than Ptolemy's.

As related by the full Latin title of the map—*Universalis Cosmographia Secundum Ptholomaei Traditionem et America Vespucii Aliorumque Lustrationes*—it was done under the influence of and in the tradition of Ptolemy and based on information supplied by Vespucci. Hence, Waldseemüller was honoring both men with prominent positions on the map. But to establish his authority and credibility, he was also consciously associating his work with their renown.

PLAN DE L'IMPORTANTE ET GLORIEUSE BATAILLE DE BLANGIES.

Anna Beeck (b. 1678, active 1693–1717) was an engraver, publisher, and colorist at The Hague in the Netherlands. It was rare for women to work in the field of cartography, and Beeck may have taken over the business from her husband. There were other female cartographers by this time and even before, but they remain largely unknown. From 1697 to 1717, Beeck issued approximately sixty spectacular, colorful maps and prints, mostly of battlefields, sieges, and fortifications, such as the composite atlas *A Collection of Plans of Fortifications and Battles 1684–1709* (1709), which was produced with several collaborators. (A composite atlas—also known as an *atlas factice*—is a loose-leaf or bound compilation of previously published maps assembled by a client or collector; such an atlas typically does not have a title page, table of contents, page numbers, or index.) The allegorical and historical figures that decoratively populate many of the maps were intended to contribute to viewers' experiences of the forts and battles shown.

One of the key battles during the War of the Spanish Succession—the Battle of Malplaquet (September 11, 1709), a strategic victory for the French—is shown on the Beeck map at left. Portraits of the victorious allied commanders—Prince Eugene of Savoy, the Duke of Marlborough, and the Comte de Tilly—are in the lower left corner. The map is named for the Blangies village or woods near Malplaquet, France. In the lower right corner of the Blangies plan, there are several allegorical figures, including a skin-clad ancient Teutonic warrior representing the age-old struggles between the Germanic and Latin peoples of Europe.

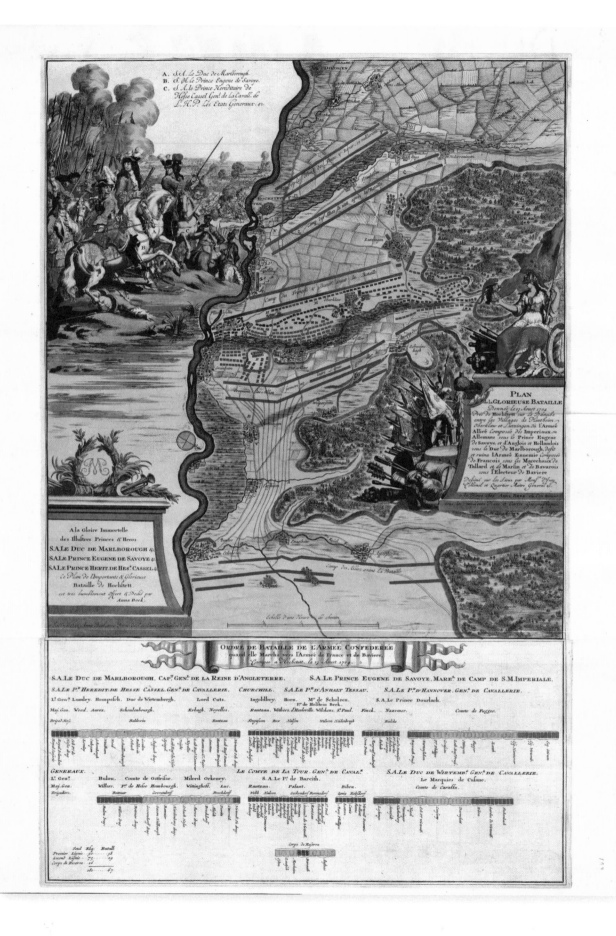

L'AMERIQUE DIVISEE EN TOUS SES PAYS ET ETATS . . . In addition to being a map, the four-sheet *L'Amerique Divisee en Tous Ses Pays et Etats . . .* (Paris, 1754), by French cartographers Sebastian G. Longchamps (1718–93) and Jean D. Janvier (active 1740s–90s), provides a French High Enlightenment history and geography of the Americas. The entire border of this large map is composed of superbly engraved vignettes of the major events of the European appropriation of the New World from the Indians and their slow "uplifting" to Christianity and civilization. In the center of the bottom margin there are two summary paragraphs on the history and geography of the Americas. The map was created from the contemporary European point of view, especially regarding the superiority of Western civilization; this was a map by Europeans for a European audience.

PLAN DE LA GLORIEUSE BATAILLE . . . PLINTHEIM . . . Contemporary personages are also depicted on the *Plan de la Glorieuse Bataille Donnee . . . Villages de Plintheim, Oberklaw et Lutzigen,* depicting another major event, the Battle of Blenheim (the Second Battle of Höchstädt), fought on August 13, 1704. In the upper left corner, the allied leaders Savoy, Marlborough, and Cassel sit proudly astride their horses, observing the battle.

THE UNITED STATES OF AMERICA . . .

On his 1783 map of America entitled *The United States of America Laid Down from the Best Authorities, Agreeable to the Peace of 1783,* the London map publisher John Wallis Sr. (1714–93) was one of the first to show the new United States as fixed by the Treaty of Paris (1783). It is a map apparently made without malice. The title cartouche in the lower right corner contains a small crowd of historical and symbolic figures, including George Washington, Benjamin Franklin, Liberty, Wisdom (Athena), and Justice. On one side, Washington is arm in arm with Liberty; on the other, Franklin is apparently drafting part of the treaty, attended by Wisdom and Justice.

PHELPS'S NATIONAL MAP OF THE UNITED STATES . . . More than a half century later, the lithographed *Phelps's National Map of the United States: A Travellers Guide,* published by Humphrey Phelps in New York in 1849, was issued in the popular portable pocket format in response to the California Gold Rush. A very nationalistic American map, it had a border composed of pictures of the U.S. presidents separated by the seals of the states of the Union at that time. It was reflective of the vitality and growth of the United States, geographically and politically.

GEOGRAPHISCHE VERBREITUNG DER MENSCHEN-RASSEN.
German geographer and engineer Dr. Heinrich Berghaus (1797–
1884) attempted a more scholarly approach for his anthropological
map entitled *Geographische Verbreitung der Menschen-Rassen,*
published in Gotha, Germany, in 1848. In order to graphically depict
the geographical spread of the human race across the world,
representatives of different groups of people—from Papuans to
Eskimos, gauchos to Hottentots—are arrayed around the top half
of the map. The portraits, some of which seem astonishing and
even monstrous today, were meant to be the product of a relatively
"scientific" dispassion. There also are two pictures comparing the
skulls of several of these groups. The map stresses geographical
influences on human diversity and development. The labeling of
the variously identified human groups in different parts of the
world strongly ties their physical and social developments, rightly
or wrongly, to geography—for example, equating darker skins and
underdevelopment to hotter climates.

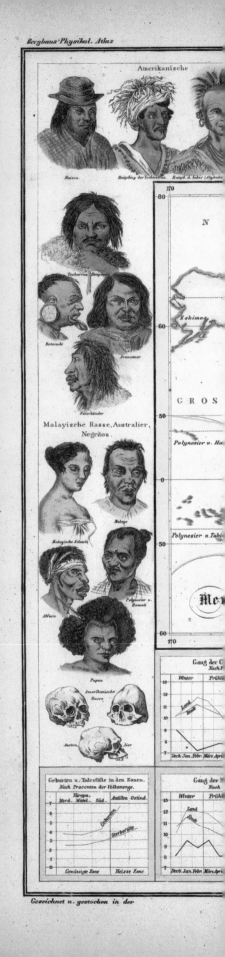

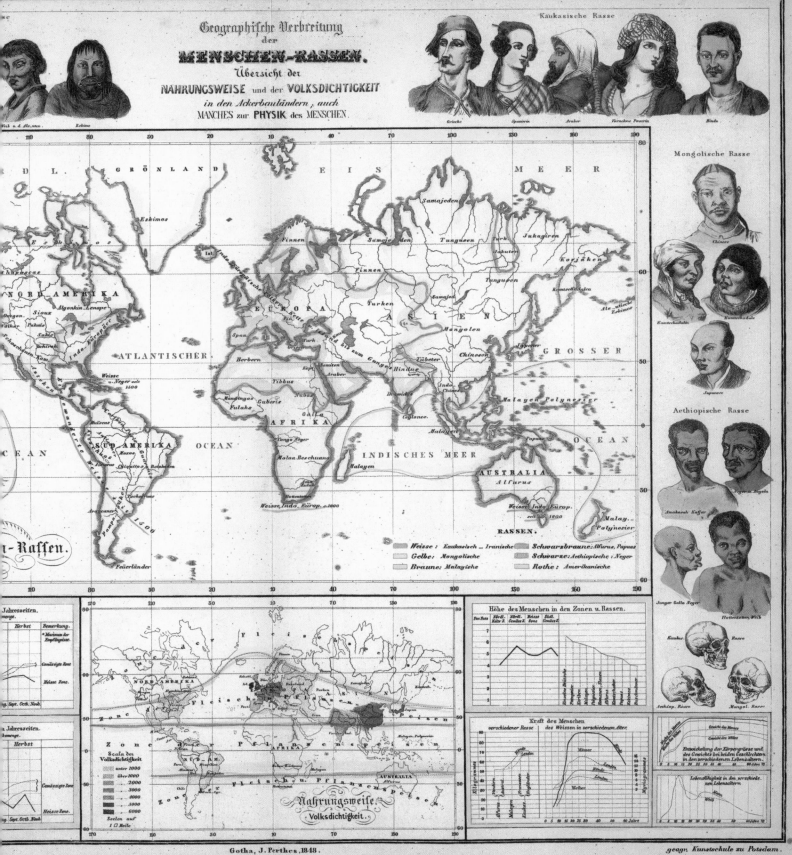

Geographische Verbreitung der
MENSCHEN-RASSEN.
Übersicht der
NAHRUNGSWEISE und der VOLKSDICHTIGKEIT
in den Ackerbauländern, auch
MANCHES zur PHYSIK des MENSCHEN

Kaukasische Rasse

Mongolische Rasse

Aethiopische Rasse

RASSEN.

Weisse : Kaukasisch — Iranische
Gelbe: Mongolische
Braune: Malayische
Schwarzbraune: Alfuren, Papuas
Schwarze: Aethiopische: Neger
Rothe : Amerikanische

Höhe des Menschen in den Zonen u. Rassen.

Kraft des Menschen
verschiedener Rasse   des Weissen in verschiedenem Alter.

Nahrungsweise.
Volksdichtigkeit.

Scala der Volksdichtigkeit
unter 1000
über 1000
2000
3000
4000
5000
6000
Seelen auf 1 ☐ Meile

Physikalischer Atlas

Heinrich Berghaus authored a number of theoretical works on mapmaking. As the map *Geographische Verbreitung der Menschen-Rassen* (previous page), part of Berghaus's monumental *Physikalischer Atlas*, exemplifies, he also was one of the founders of scientific thematic cartography: the production of maps that explore and illustrate specific geographic topics. The *Physikalischer Atlas* was published in two editions, 1838–48 and 1849–52. It was Intended to accompany the early volumes of *Kosmos* (1845–58), the magnum opus of the virtuoso German polymath intellectual—naturalist, explorer, author, philosopher, and diplomat—Alexander von Humboldt (1769–1859).

Shown below is a detail from Geographical Division and Distribution of Reptilia (Reptiles), *one of the zoological geography maps in* The Physical Atlas: A Series of Maps and Notes Illustrating the Geographical Distribution of Natural Phenomena, by Alexander Keith Johnston . . . based on the Physikalischer Atlas of Professor H. Berghaus, *published by William Blackwood and Sons in London in 1848. On the opposite page is a detail of another map from the atlas entitled* Botanical Geography of the World.

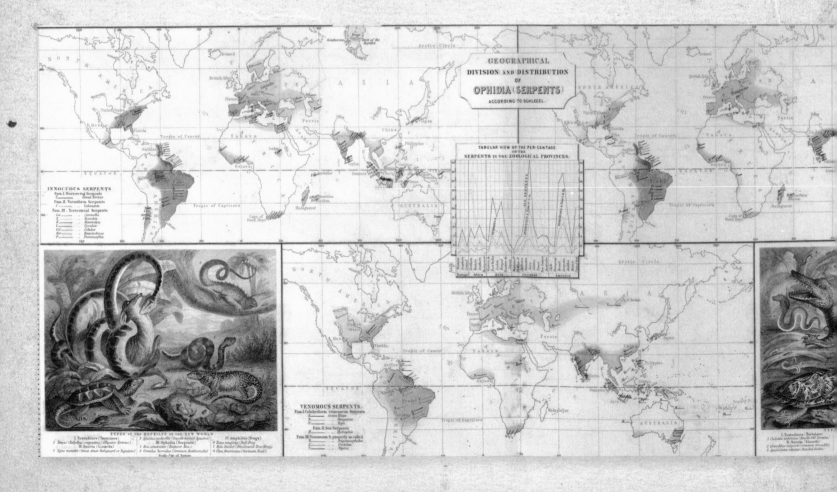

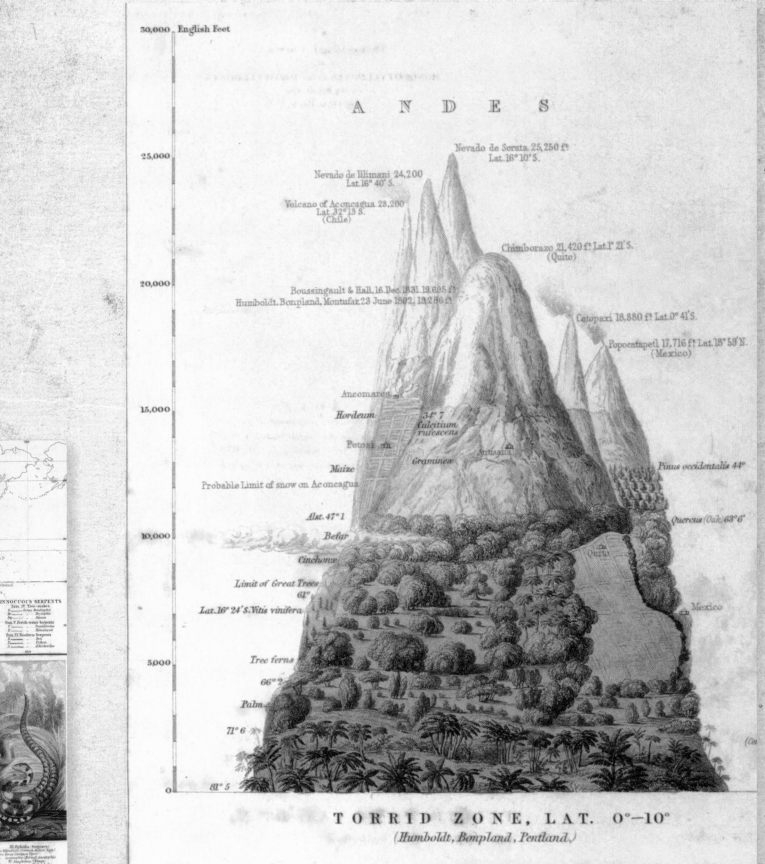

ANDES

30,000 English Feet

Nevado de Sorata 25,250 f.t
Lat. 16° 10' S.

Nevado de Illimani 24,200
Lat. 16° 40' S.

25,000

Volcano of Aconcagua 23,200
Lat. 32° 13 S.
(Chile)

Chimborazo 21,420 f.t Lat. 1° 21' S.
(Quito)

20,000

Boussingault & Hall, 16. Dec. 1831. 19,695 f.t
Humboldt, Bonpland, Montufar, 23 June 1802, 19,286 f.t

Cotopaxi 18,880 f.t Lat. 0° 41' S.

Popocatapetl 17,716 f.t Lat. 18° 59' N.
(Mexico)

Ancomarca

Hordeum

34° 7
fulcitium
rufescens

15,000

Potosi

Antisana

Pinus occidentalis 44°

Graminex

Maize

Probable Limit of snow on Aconcagua

Alst. 47° 1

Quercus (Oak) 63° 6'

10,000

Befar

Quito

Cinchonæ

Limit of Great Trees
61°

Mexico

Lat. 16° 24' S. Vitis vinifera

Tree ferns

5,000

66° 2

Palm

71° 6

81° 5

0

TORRID ZONE, LAT. 0°—10°
(Humboldt, Bonpland, Pentland.)

Imperialism

With the full realization of the era of the highly nationalistic Western "new imperialism" in the later part of the nineteenth century, some of the cartography of the period followed suit. Historically, cartography has gone hand in hand with the development of empire and been engaged in a symbiotic relationship with it; simply put, maps have acted as necessary tools of imperialism.

BRITISH EMPIRE . . . On John Bartholomew's thematic steel-engraving map entitled *British Empire Throughout the World Exhibited in One View* (Edinburgh, c. 1850), the far-flung parts of the British empire are distinctively highlighted in red. At the top and the bottom of the map, peoples from across the empire are proudly paraded, intermingled with British citizens. The meaning is clear: British civilization and culture, spread across the globe throughout the empire, were positive forces with benefits not only for the British, but for the subject peoples of the empire as well. The designer of the map, John Bartholomew Jr. (1831–93), eventually took over the leadership of the Edinburgh map-publishing firm John Bartholomew and Company (c. 1826–1960s) from his father, John Bartholomew Sr. (1805–61). In the 1830s, the company received the contract for making maps of this type for the *Encyclopedia Britannica.*

Natives of British Asia and of East Indian Islands.

A. FULLARTON & Cº.

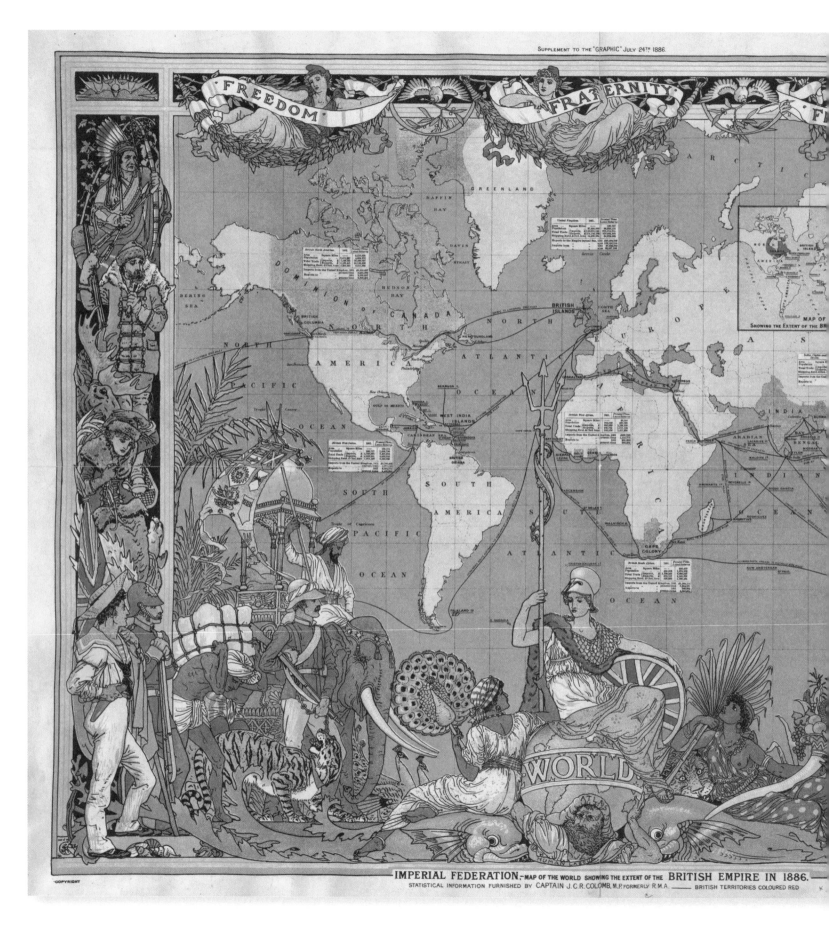

FREEDOM · FRATERNITY · F

IMPERIAL FEDERATION,—MAP OF THE WORLD SHOWING THE EXTENT OF THE BRITISH EMPIRE IN 1886.

STATISTICAL INFORMATION FURNISHED BY CAPTAIN J.C.R. COLOMB, M.P. FORMERLY R.M.A. ——— BRITISH TERRITORIES COLOURED RED

'COPYRIGHT

IMPERIAL FEDERATION . . . J. C. R. Colombo's *Imperial Federation—Map of the World Showing the Extent of the British Empire in 1886* is unquestionably the epitome of the cartographic expression of British new imperialism at its height. This elegantly designed and beautifully colored map appeared as a supplement to the *Graphic,* an illustrated British newspaper, on July 24, 1886. Once again, the parts of the empire are illustrated in red.

This world map is centered on what was established in 1884 as the internationally agreed upon prime meridian for all cartography, running north to south through Greenwich, England, the site of the Royal Observatory, which puts the United Kingdom just above the main focal point of the map. The map is based on the Mercator Projection, whose distortions exaggerated the extent of Canada in North America. The Mercator Projection was devised by the Flemish cartographer Gerardus Mercator (1512–94), who first applied it to his great world map published in Duisburg in the Rhineland in 1569. In this cylindrical projection, the parallels increase in spacing with greater distance from the equator, and large objects (e.g., Greenland) are distorted significantly, but it is of major help to navigation because all straight lines (i.e., courses) on a Mercator Projection map are constant compass directions.

Around the left, bottom, and right sides of Colombo's map, the peoples and beasts (elephant, tiger, kangaroo) of the empire are spectacularly arrayed around the figure of Britannia, seated at the bottom center of the map. The human figures, representing many different countries and cultures, are meant to symbolize the diverse, cooperative, peaceful family of peoples of the Britannic "federation." Their prominent placement on the map touts this global community as one of the significant outcomes of British civilization. But the main message of the map, because it was issued by a leading British publication, was directed at Great Britain's citizenry at home and abroad rather than the other peoples of the empire who grace the map's borders: The British were to be proud of their global imperial achievements, but they must also accept the enormous costs of empire.

A   C I F I Q U E   Cop

I. Rapa   la Seren

St. Michel   Valparais   Santia

I. Juan Fernandez

Constituti

la Conception

Valdivia

I. Chiloé

Archipel

Chonos

I. Wellington

I. de la Rne. Adelaide

Tre. de la Désolation

1821 pa

7

Of Human Achievements...
Conclusion

"Our age today is doing things of which antiquity did not dream. . . . A new globe has been given us by the navigators of our time."

—Jean Fernel, Dialogue, 1530[41]

Not only did cartographers enhance their works with likenesses of humanity; they also freely added illustrations of major monuments, important edifices, and other notable constructions. As with many cartographic embellishments, these pictures functioned beyond just filling space or merely beautifying maps; they were also intended to improve the total cartographic experiences of map viewers as well as to increase the marketability of maps. Site-specific imagery brought greater clarity by visually enhancing the two-dimensional representations of places depicted on maps. Plants, animals, people, ships, and buildings helped map users mentally travel between the abstractions of cartographic symbolization and line and the actualities of the places and actions shown.

Military Structures

Such pictography—expanded to include armaments, batteries, personnel formations, and battle strategies—was often used to illustrate the placement of fortifications and the movement of troops on military maps, both simple and elaborate.

PREVIOUS PAGES: *Right*, **GRAND BIRDS EYE VIEW OF THE GREAT EAST RIVER SUSPENSION BRIDGE, DETAIL.** Currier & Ives, . . . New York, c. 1885. *Left*, **PLANISPHERE, VIGNETTE.** L. Smith for L. E. Desbuissons and J. Migeon, Paris, 1892.

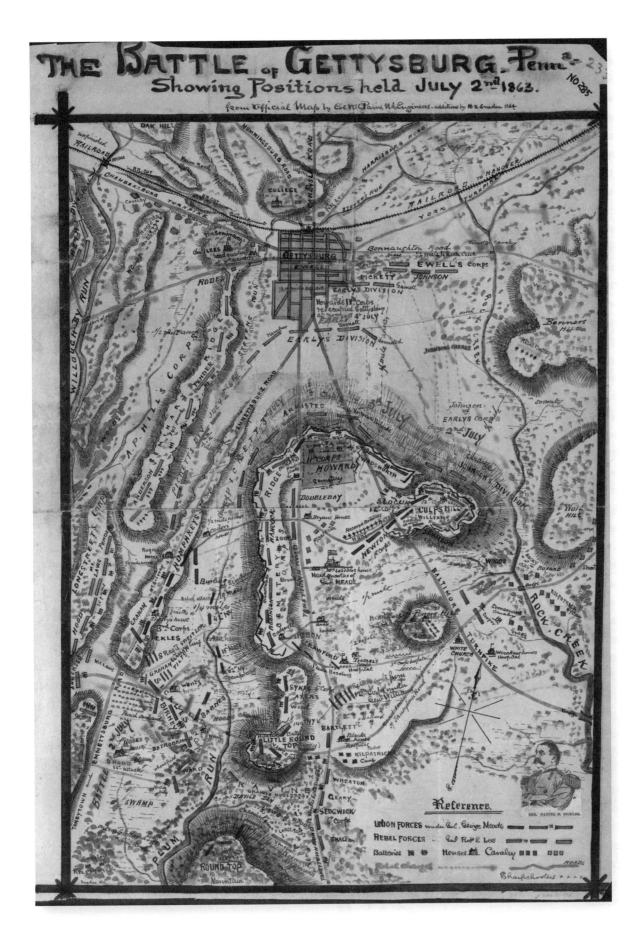

THE BATTLE OF GETTYSBURG. PENNA. SHOWING POSITIONS HELD JULY 2ND, 1863. This map, by prolific Civil War artist and Union Army cartographer Private Robert Knox Sneden, shows the locations of specific units during the Battle of Gettysburg. Red arrows indicate movement, such as the direction of charges and assaults. Landmarks such as Culp's Hill, Cemetery Ridge, and Little Round Top, are indicated, as well as the locations of hospitals and headquarters. Color coding indicates the location of Union (purple) and Confederate (red) forces.

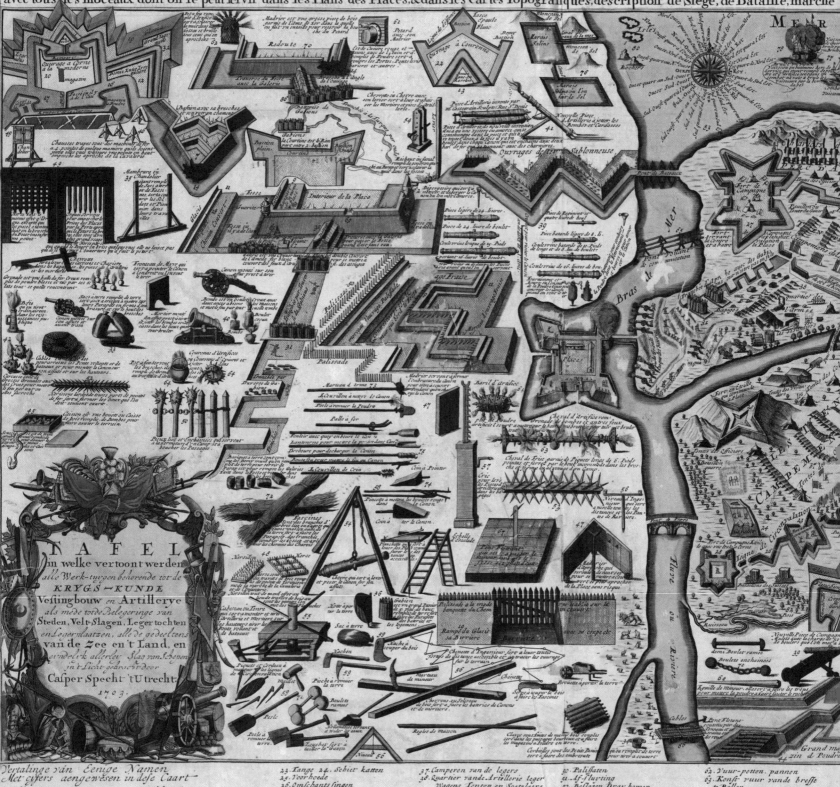

CARTE QUIREPRESENTE TOUTES LES PIECES QUI SONT COMPRISES DANS L'ARCHITECTURE MILITA
avec tous les mo̅ceaux dont on se peut servir dans les Plans des Places, & dans les Cartes Topografiques, description de Siege, de Bataille, marche

TAFEL
in welke vertoont werden
alle Werk-tuygen behoorende tot de
KRYGS-KUNDE
Vestingbouw en Artillerye
als mede wide Belegeringe van
Steden, Velt-Slagen, Leger tochten
en Legerplaatzen, alle de gedeeltens
van de Zee en 't Land, en
eindelyk allerley Slag van Schepen
in 't Licht gebracht Door
Casper Specht t'Utrecht.
1703.

MER

TAFEL IN WELKE VERTOONT . . . A good example of the military map genre is a beautifully detailed map entitled *Tafel in Welke Vertoont Warden alle Werktuygen Behorende tot fe Krygs Kunde . . .* (Utrecht, 1703), by Dutch engraver Casper Specht (active 1684–1708), which was included in Anna Beeck's *A Collection of Plans of Fortifications and Battles 1684–1709.* On the right side of the map are layouts of generic fortifications, built to defend against an attack from the sea. The left side shows numerous details of fortification construction, as well as the armaments needed to defend such positions: walls, moats, and the tools for their erection are richly illustrated in detail and color, as are cannons, mortars, grenades, and other weapons. In addition to serving as an in-depth explanation of defensive warfare, this initial map in the collection, with its almost encyclopedic array of matériel, was also intended to facilitate a greater understanding of the 116 other fortification and battle plans in the Beeck composite atlas. Today, this sheet is a rare gift, a valuable source for military historians on late-seventeenth-century naval siege warfare and defense strategy.

Monuments of Egypt

Besides fortifications, other ancient and modern structures were often pictured on cartography well into the nineteenth century. The Great Sphinx of Giza was a subject of fascination for mapmakers beginning with the early days of cartography. For example, on the 1572 Braun and Hogenberg map of Cairo (see full map on page 141), from their *Civitates Orbis Terrarum* atlas, the Sphinx appears in the desert in the lower right corner, along with the pyramids.

EGYPT AND ARABIA PETRAEA. At the bottom of English cartographer John Tallis's map of Egypt (1851)—part of his very popular *Illustrated Atlas*—the Sphinx is located on the Nile's east bank, as a vignette in what is today the southernmost region of Upper Egypt; an image of the ruins of the great temple at Karnak is situated directly opposite the Sphinx, on the west bank of the Nile. These representations, far to the south of the locations of the actual monuments, were positioned to fill in some of the blank spaces in the eastern Sahara Desert surrounding the Nile. They are intended to remind the map's viewers of the glory of the original Bronze Age civilization of ancient Egypt.

This theme is further carried out in the stylized border framing the map. The border begins at the top center, emanating from the sarcophagus-like title cartouche, guarded by two ram-headed sphinxes, and proceeds down both sides of the maps with motifs comprised of abstracted lotuses and papyrus plants. Sitting at the top left and top right are statues of a figure likely meant to represent Horus, the falcon-headed Egyptian god of the sky and light. The border is anchored in each lower corner by an obelisk inscribed with hieroglyphics; under the obelisk is a colossal statue of a pharaoh, reminiscent of those at the famous temple of Abu Simbel in Upper Egypt. The two sides of the border rejoin at the bottom center of the map in a splendid symbolization of Horus.

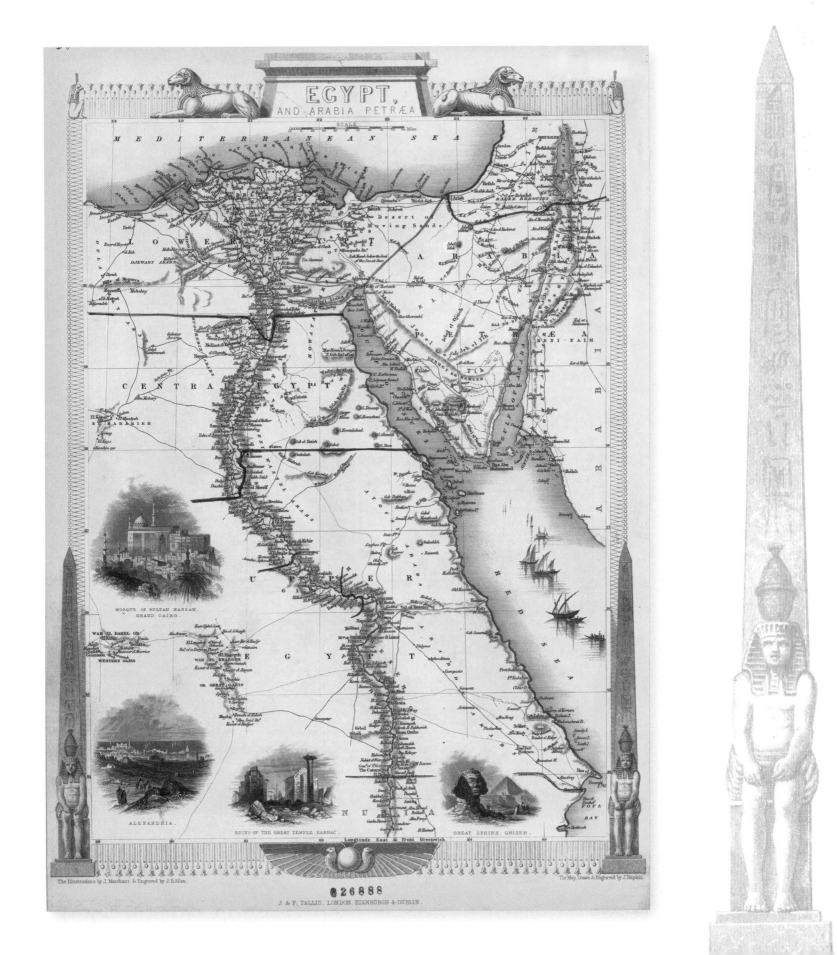

EGYPT,
AND ARABIA PETRÆA.

SCALE

MEDITERRANEAN SEA

LOWER EGYPT

CENTRAL EGYPT

ARABIA

UPPER EGYPT

RED SEA

NUBIA

MOSQUE OF SULTAN HASSAN. GRAND CAIRO.

ALEXANDRIA.

RUINS OF THE GREAT TEMPLE, KARNAC.

GREAT SPHINX, GHIZEH.

Longitude East from Greenwich

The Illustrations by J. Marchant & Engraved by J.B. Allen.

The Map, Drawn & Engraved by J. Rapkin.

J. & F. TALLIS, LONDON, EDINBURGH & DUBLIN.

The Jewel of the British Empire

There were other reasons for putting ancient regalia on maps of modern Egypt beyond the covering up of empty space and decoration. In 1856, the Egyptian ruler Sa'id Pasha—generally considered a weak ruler manipulated by the imperialists—granted the contract to build a canal joining the Mediterranean and Red seas to the French developer Ferdinand de Lesseps. The building of the canal marked the beginning of European penetration (following Napoleon's failed occupation in 1798–1801) of an Egypt only loosely controlled by the Ottoman Empire. Great Britain originally opposed the French construction of the canal—which was opened to shipping in November 1869—but by 1875 it had bought out Egypt's interests and become the canal's controlling shareholder. Thereafter, the Suez Canal became the gateway to India, the "jewel of the British Empire," and Egypt was forced into a sham alliance with Britain. Ostensibly to finally end slavery and to put down the religious revolt of the Islamic Mahdists (followers of a Muslim cleric named Muhammad Ahmad) in neighboring Sudan, Britain by 1899 had extended its influence even farther southward up the Nile into the Sudan to establish the condominium (jointly ruled government) of Anglo-Egyptian Sudan. When the Ottoman Empire joined the Central Powers during World War I, Britain declared Egypt a British protectorate. Egypt finally gained its independence from Britain in 1923; Sudan remained an Anglo-Egyptian condominium until 1956.

In the nineteenth century, English maps of Egypt like the one by Tallis were reflective of the era of Europe's "new imperialism," which began with the First Opium War (1839–42), a trade war fought between Britain and China. In relating modern Egypt back to the splendor of its ancient past, such maps helped persuade British viewers of the importance of Egypt as a possible territory. Public opinion was important, for empires were expensive to maintain; consequently, taxes had to help pay for them. In the nationalistic nineteenth century, pride of ownership—knowing that their countries now exercised sway over once-great nations—made the economic and social burdens somewhat easier to bear for the Europeans who had to pay the price of empire.

This illustration depicts one of the inaugural ceremonies surrounding the inauguration of the Suez Canal. It appeared in the January 1, 1870, edition of *Frank Leslie's Illustrated Newspaper,* a weekly American magazine.

New World Feats

Beyond the antique marvels of the Old World, the modern wonders of the New World were publicized on maps as well. The Mavericks were a family of silversmiths, engravers, and printers who had a thriving business in Newark, New Jersey, and then New York City in the first half of the nineteenth century. In 1833, a second-generation member of the family, Samuel Maverick (1789–1845), published a pocket *Map of the State of New York* for travelers, especially those moving westward. Chief among the routes to be followed across New York on this map is the system of the Grand Canal, also called the Erie Canal. Officially opened in 1825, it eventually was extended some 360 miles across the central part of the state between Albany and Buffalo, connecting the Hudson River to Lake Erie. An engineering marvel, the canal was the pride of the new United States. It contributed to national unity, provided easier and less expensive transportation, and facilitated increased trade. Only in the 1850s did competition from railroads begin to lessen the canal's long-haul advantages, but it continued to be a profitable route into the 1880s.

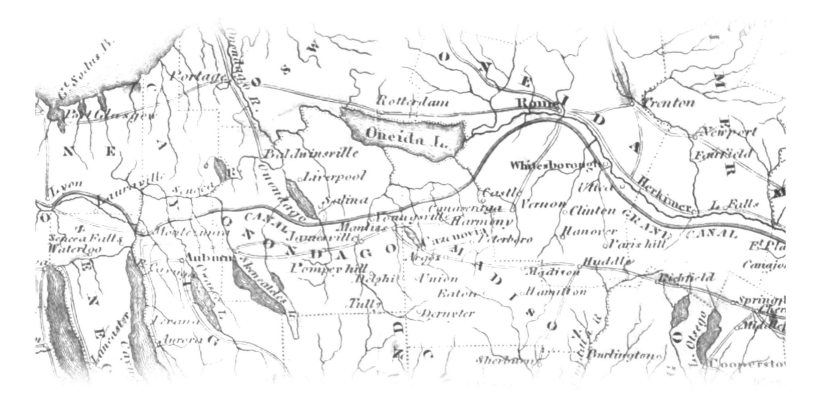

GRAND CANAL

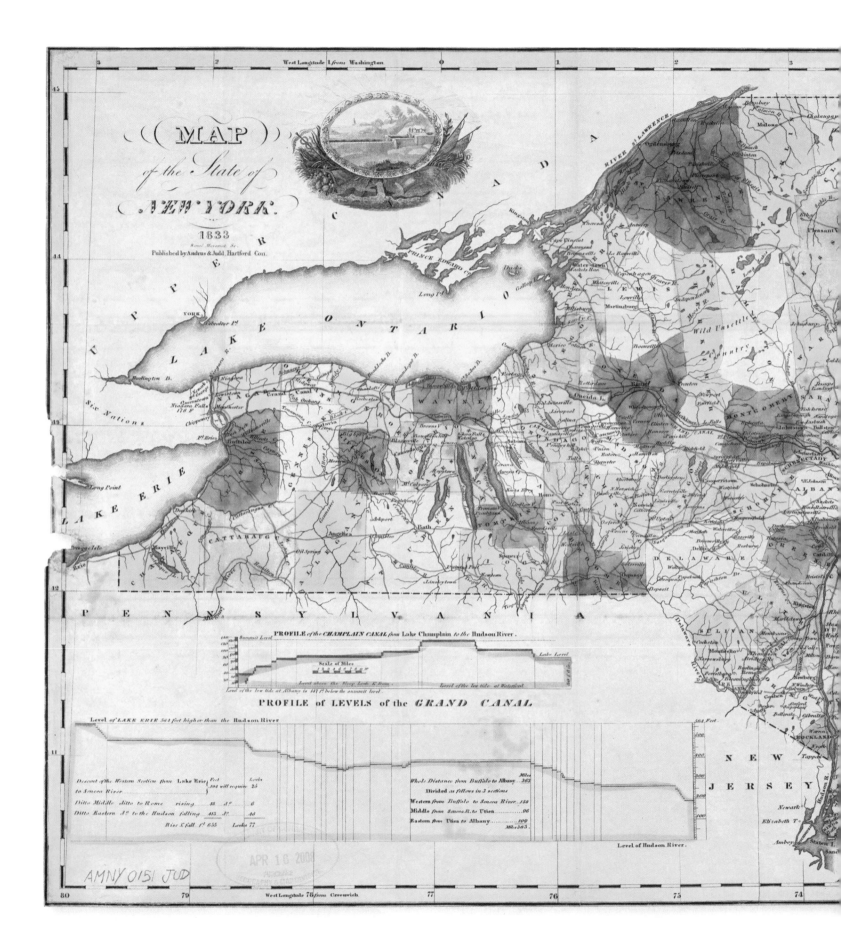

((MAP))

of the State of

NEW YORK.

1833

Saml Maverick Sc.

Published by Andrus & Judd, Hartford Con.

PROFILE of the CHAMPLAIN CANAL from Lake Champlain to the Hudson River.

Summit Level

Lake Level

Scale of Miles

Level above the Sleep Lock E Dam.

Level of the low tide at Waterford

Level of the low tide at Albany is 147 f below the summit level.

PROFILE of LEVELS of the GRAND CANAL

Level of LAKE ERIE 564 feet higher than the Hudson River

564 Feet

Descent of the Western Section from Lake Erie	Feet	Locks
to Seneca River	} 104 will require	25
Ditto Middle ditto to Rome rising	48 d°	6
Ditto Eastern d° to the Hudson falling	413 d°	46
Rise & fall f. 655	Locks	77

Whole Distance from Buffalo to Albany	Miles
	363
Divided as follows in 3 sections	
Western from Buffalo to Seneca River	158
Middle from Seneca R. to Utica	96
Eastern from Utica to Albany	109
	Mls. 363

Level of Hudson River.

MAP OF THE STATE OF NEW YORK. Maverick's map extolled the Erie Canal. The lower left of the map is dominated by two diagrams. The upper one, "Profile of the Champlain Canal from Lake Champlain to the Hudson River," shows the levels of the Champlain Canal, another construction that opened in 1819 and that connected Lake Champlain to the Hudson River. The bottom image, "Profile of Levels of the Grand Canal," covers the entire Erie Canal and clearly demonstrates how much higher the level of Lake Erie is than that of the Hudson River. Just to the right of the title there is a picture of two men escorting a mule-drawn barge through one of the canal's thirty-six locks (see *above*). The locks, with their varying levels, were a major part of the canal's achievement, for they allowed commerce to efficiently traverse the system. Maverick's map is almost proudly immodest in documenting an important historical chapter in the rise of the American nation.

HOLT'S NEW MAP OF WYOMING . . . , VIGNETTE. A similar type of braggadocio informs Holt's previously discussed 1883 map of the Wyoming Territory (see full map on page 124). While factories in the 1880s were still scarce in Wyoming, they were coming to command much of the American landscape elsewhere. The reassuring image of the vibrant factory to the right of the title thus proclaims not only the potential value of Wyoming as a would-be state but the might of the American nation as it began assuming leadership of the Industrial Revolution.

PREVIOUS PAGES: THE CITY OF NEW YORK . . .
In 1883, Currier & Ives published this bird's-eye-
view map of Manhattan, celebrating the marvel
that was the Equitable Life Assurance Building.
Construction was completed in 1870, and it was the
first office building to include passenger elevators.
At a then-record 130 feet, it was the tallest non-
religious building in the world—a position it held
until 1884—and is considered by many to be the
world's first skyscraper (for comparison, the Empire
State building, without its antenna, is 1,250 feet).
The Equitable building is shown in white in the left
foreground, festooned with a flag emblazed with
its name (enlarged *above*).

An Invention of Vast Use

It is apparent that pieces of imagery that marked
human accomplishments were important elements of maps for a number of interrelated
reasons. They provided a further form of expression for mapmakers, illustrators, and designers.
They were intended to broaden the compre-
hension of map users with regard to the specific
geography depicted. They often exuded pride of
place—whether nationalistic or imperialistic—and
exalted human exploits, past and present. These
illustrations also regularly enhanced the attractiveness (and salability) of commercial and even
official cartography.

But there may be more to be grasped here, as
suggested in the quotation at the beginning of this
chapter from the French Renaissance physician,
cartographer, astronomer, and mathematician
Jean Fernel, and the one below from the eminent
British Enlightenment geographer and cartographer Herman Moll. Although these learned
mapmakers lived almost two centuries apart, both
men clearly recognized that maps themselves
were—and continue to be—significant exemplars
and monuments of human attainment.

Detail of an advertisement from Thomas Jefferys, royal engraver and geographer to the Prince of Wales, c. 1750, featuring a muse holding and surrounded by the instruments of cartography.

"The art of making of maps and sea-charts is an invention of such vast use to mankind that perhaps there is nothing for which the World is more indebted to studious Labours of Ingenious Men."
—Herman Moll, Atlas Manuale, 1709[42]

NICÆA

CIVITAS.

1. David Greenhood, *Mapping* (Chicago and London: University of Chicago Press, 1964), 13.

2. Charles Close, *The Map of England, or About England with an Ordnance Map* (London: Peter Davies, 1932), 13.

3. Norman J. W. Thrower, *Maps and Civilization: Cartography in Culture and Society,* third ed. (Chicago and London: University of Chicago Press, 2007 [1972]), 51.

4. Robert Temple, *The Genius of China: 3,000 Years of Science, Discovery, and Invention* (London: Prion, 1999 [1986]), 148–57.

5. C. L. Verbaek and S. Thirslund, *The Viking Compass Guided Norseman First to America* (Skjern, Denmark, 1992), 149.

6. Thrower, 51–56.

7. Jill Lisk, *The Struggle for Supremacy in the Baltic: 1600–1725* (New York: Funk and Wagnalls, 1967), 17–18.

8. Dennis Reinhartz, *The Cartographer and the Literati: Herman Moll and His Intellectual Circle* (Lewiston, NY/ Lampeter, Wales: Edwin Mellen Press, 1997), 113–49.

9. Katherine R. Goodwin and Dennis Reinhartz, *Tabula Terra Nova* (Dallas, TX: Somesuch Press, 1992), 1–2.

10. Robert Herrick, "A Country Life: To the Honoured M. End Porter Groom of the Bedchamber to His Majesty," *Works of Robert Herrick,* vol. II (London: Lawrence and Bullen, 1891), 33–35.

11. Dennis Reinhartz, "The Americas Revealed in the *Theatrum,*" in *Abraham Ortelius and the First Atlas: Essays Commemorating the Quadricentennial of His Death, 1598–1998,* ed. by Marcel van den Broecke, Peter van der Krogt, and Peter Meurer (Utrecht, Netherlands: HES Publishers, 1998), 209–20.

12. James A. Welu, "The Sources and Development of Cartographic Ornamentation in the Netherlands," in *Art and Cartography: Six Historical Essays,* ed. by David Woodward (Chicago and London: University of Chicago Press, 1987), 147–73.

13. Philip D. Burden, *The Mapping of North America: A List of Printed Maps, 1511–1670,* 2 vols. (Rickmansworth, UK: Raleigh, 1996), xiv–xvi.

14. Samuel Taylor Coleridge, from *Rime of the Ancient Mariner,* 1798.

15. For example, see William A. R. Richardson, *Was Australia Charted before 1606?: The Java la Grande Inscriptions* (Canberra: National Library of Australia, 2006) and Peter Trickett, *Beyond Capricorn: How Portuguese Adventurers Secretly Discovered and Mapped Australia and New Zealand 250 Years Before Captain Cook* (Bowden, S.A., Australia: East Street Publications, 2007).

16. Tony Campbell, *Early Maps* (New York: Abbeville, 1981), 88–89.

17. Ashley and Miles Baynton-Williams, *New Worlds: Maps from the Age of Discovery* (London: Quercus, 2006), 42–43.

18. Sir H. G. Fordham, *Some Notable Surveyors and Map-Makers of the Sixteenth, Seventeenth and Eighteenth Centuries, and Their Work* (Oxford: Clarendon Press, 1929).

19. R. J. Majors (ed.), *Select Letters of Christopher Columbus: With Other Original Documents, Relating to His Four Voyages to the New World* (New York: Corinth Books, 1961), 26–27.

20. Daniel Defoe, *A Tour Thro' the Whole Island of Great Britain, Divided into Circuits or Journies . . . by a Gentleman,* 3 vols. (London: 1724–27).

21. Peter Earle, *The World of Defoe* (London: Wiedenfeld and Nicolson, 1976), 39.

22. Toby Lester, *The Fourth Part of the World: The Race to the Ends of the Earth, and the Epic Story of the Map That Gave America Its Name* (New York: Free Press, 2009), 50–53.

23. Alvar Nuñez Cabeza de Vaca, *Castaways: The Narrative of Alvar Nuñez Cabeza de Vaca,* ed. by Enrique Pupo-Walker (Berkeley and Los Angeles: University of California Press, 1993), 63.

24. Katherine R. Goodwin and Dennis Reinhartz, *Tabula Terra Nova.*

25. Edward H. Dahl, "The Original Beaver Map— De Fer's 1698 Wall Map of America," *The Map Collector* (December 1984), 22–26.

26. Jonathan Swift, *Gulliver's Travels* (New York: W. W. Norton, 1970 [1726]), 103–116.

27. Reinhartz, *The Cartographer and the Literati,* 37 and 133–144.

28. Gillian Hill, *Cartographical Curiosities* (London: British Library, 1978), 38–41.

29. Tony Campbell, *Early Maps* (New York: Abbeville, 1981), 94–95; H.A.M. van der Heijden, *Leo Belgicus: An Illustrated and Annotated Carto-bibliography* (Alphen aan den Rijn: Canaletto, 1990), 6–16; and Hill, 39–43.

30. Van der Heijden, 19 and 37–42.

31. Charles Baudelaire, "The Voyage," *The Flowers of Evil,* 1857.

32. Páll Bergþórsson, *The Wineland Millennium: Saga and Evidence* (Reykjavik: Millennium Commission of Iceland, 2000) and Gwynn Jones, *The Norse Atlantic Sagas: Being the Norse Voyages of Discovery and Settlement to Iceland, Greenland, and North America,* second ed. (Oxford and New York: Oxford University Press, 1986 [1964]).

33. John Block Friedman, *The Monstrous Races in Medieval Art and Thought* (Cambridge, MA, and London: Harvard University Press, 1981) and Lester, 37–41.

34. Dennis Reinhartz, "Enlightenment Maps and the "Noble Savage," *IMCoS Journal 114* (Autumn 2008), 39–46.

35. Thomas Harriot, "A Brief and True Report on the New Found Land of Virginia," in *India Occidentalis,* ed. by Theodore de Bry (Frankfurt am Main: 1590).

36. Dennis Reinhartz, "Establishing a Transatlantic Graphic Dialogue, 1492–1800," in *Transatlantic History,* ed. by Stephen G. Reinhardt and Dennis Reinhartz (College Station: Texas A&M University Press, 2006), 53–54. Also see Michael Gaudio, *Engraving the Savage: The New World and Techniques of Civilization* (Minneapolis and London: University of Minnesota Press, 2008).

37. Alexander Pope, *An Essay on Man,* 1732–34.

38. Robert F. Berkhofer, *The White Man's Indian: Images of the American Indian from Columbus to the Present* (New York: Knopf, 1978), 43–44 and 74–77.

39. David Weber, *Bárbaros: Spaniards and Their Savages in the Age of the Enlightenment* (New Haven, CT: Yale University Press, 2005), 19 and 43–44.

40. Dennis Reinhartz, *The Cartographer and the Literati,* 97–142.

41. Jean Fernel, *Dialogue* (Paris: 1530). In 1525 Fernel (c. 1497–1558) drew a map of France and measured the meridian between Paris and Amiens.

42. Herman Moll, *Atlas Manuale,* 1709.

PORTOLAN ATLAS OF NINE CHARTS AND A WORLD MAP . . . ,
BACK COVER. Battista Agnese, Venice, 1544.

CARTOBIBLIOGRAPHY

Agnese, Battista. *Portolan Atlas of Nine Charts and a World Map* . . . Venice: 1544.

——. *Portolan Atlas* . . . back cover. Venice: 1544.

Aitzing, Baron Michael von, and Frans Hogenberg. *Leo Belgicus*. Cologne: 1583.

d'Anville, Jean Baptiste Bourguigon. *Province de Quang-Tong*, from *Nouvel Atlas de la Chine*. Paris and The Hague: Henri Scheurleer, 1737.

Austin, Stephen F. *Map of Texas with Parts of the Adjoining States*. Philadelphia: Henry S. Tanner, 1830.

Bachmann, John. *Birds Eye View of Louisiana, Mississippi, Alabama, and Part of Florida*. New York: 1861.

Bartholomew, John, Jr. *British Empire throughout the World Exhibited in One View*. Edinburgh: c. 1850.

Beeck, Anna. *A Collection of Plans of Fortifications and Battles* [atlas factice]. The Hague: 1684–1709.

A Collection of Plans . . . , *Plan de la Glorieuse Bataille Donnee . . . Villages de Plintheim, Oberklaw et Lutzigen*. The Hague: 1704.

A Collection of Plans . . . , *Plan de l'Importante et Glorieuse Battaille de Blangies*. The Hague: 1709.

A Collection of Plans . . . , *Nicea Civitas*. The Hague?: c. 1684–1709.

A Collection of Plans . . . , *Théâtre de la Paix entre les Chrétiens et les Turcs*. The Hague?: c. 1684–1709.

Berghaus, Heinrich. *Geographische Verbreitung der Menschen-Rassen*. Gotha, Germany: 1848.

——. *The Physical Atlas* . . . *Based on the Physikalischer Atlas of Professor H. Berghaus*, by Alexander Keith Johnston. Edinburgh and London: William Blackwood & Sons, 1848 [original Berghaus edition, Gotha: 1838–48].

The Physical Atlas . . . , *Botanical Geography of the World*. Johnston. Edinburgh and London: William Blackwood & Sons, 1848.

The Physical Atlas . . . , *Survey . . . of the Most Important Plants Which Are Used as Food for Man*. Johnston. Edinburgh and London: William Blackwood & Sons, 1848.

The Physical Atlas . . . , *Zoological Geography of the World*. Johnston. Edinburgh and London: William Blackwood & Sons, 1848.

Blaeu, Joan and Cornelis. *Novam Hanc Territorii Francofvrtensis Tabulam* . . . Amsterdam: c. 1638.

Blaeu, Willem Janszoon. *Africae nova descriptio*. Amsterdam: 1630.

——. *Rvssiae* . . . , from *Le Theatre du Monde, ou, Novvel Atlas*. Amsterdam: 1647.

Boazio, Giovanni Baptista. *Irlandiae accvrata descriptio*. Antwerp: Abraham Ortelius [*Theatrum Orbis Terrarum*], 1606.

——. Map and views illustrating Sir Francis Drake's West Indian voyage, from *A Summarie and True Discourse of Sir Francis Drake's West Indian Voyage*. London: Biggs and Croftes, 1588–89.

A Summarie and True Discourse, *Cartagena*, London: Biggs and Croftes, 1588–89.

A Summarie and True Discourse, *Santiago, Cape Verde*. London: Biggs and Croftes, 1588–89.

A Summarie and True Discourse, *Santo Domingo*. London: Biggs and Croftes, 1588–89

Boileau de Bouillon, Gilles. *Septemtrionalivm Regionvm Svetiae Gothiae Norvegiae Daniae* . . . from *Geografia Tavole Moderne di Geografia*. Rome: Antoine Lafréry, c. 1575.

Braun, George and Frans Hogenberg. *Civitates Orbis Terrarum* [atlas]. Cologne: c. 1576 [1572–1617].

    Civitates Orbis Terrarum, Byzantium nunc Constantinopolis. Cologne: 1572–1617.

    Civitates Orbis Terrarum, Cairos qvae olim Babylon; Aegypti Maxima Vrbs. Cologne: 1593 [1572–1617].

    Civitates Orbis Terrarum, Mexico, Regia et Celebris Hispaniae Novae Civites / Cvsco Regni Perv in Novo orbe Capvt. Cologne: 1582 and 1593 [1572–1617].

Bunting, Heinrich. *Asia Secunda Pars Terrae in Forma Pegasi*. Hanover, Germany: 1581.

Carte de la Partie de la Virginie ou l'Armée Combinée de France & des États-Unis . . . York-town & de Glocester. Paris: Esnauts et Rapilly, 1781.

Cellarius, Andreas. *Hemispherium Stelatum Australe*, from *Harmonia Macrocosmica, seu, Atlas Iniversalis et Novus . . .* Amsterdam: 1708.

Châtelain, Henri-Abraham. *Carte Tres Curieuse de la Mer du Sud . . .* Amsterdam: 1719.

Cock, Hieronymous. *Americae Sive Qvartae Orbis Partis Nova Exactissima Descriptio . . .* Antwerp: 1562.

Colombo, J. C. R. *Imperial Federation—Map of the World Showing the Extent of the British Empire in 1866*. London: 1886.

Colton, J. H. *Map of the United States of America*. New York: 1850.

Colton, J. H. and A. J. Johnson. *Johnson's New Illustrated Family Atlas . . .* New York: 1862.

Cóvens, Jean. *Archipelague du Mexique ou Sont les Isles de Cuba, Espagnole, Iamaïque, &c*. Amsterdam: c. 1757.

Cresques, Abraham. *Atlas Catalán* [panel 6]. Barcelona: 1375.

Currier & Ives. *The City of New York—Showing the Building of the Equitable Life Assurance Society . . .* New York: 1883.

——. *Grand Birds Eye View of the Great East River Suspension Bridge . . .* New York: c. 1885

Delisle, Guillaume. *Carte du Canada ou de la Nouvelle France . . .* Paris: 1708 [1703].

Eckebrecht, Phillipe. *Noua Orbis Terrarium Delineatio . . .* Nuremberg: 1630.

Escandón, José de. *Mapa de la Sierra Gorda y Costa del Seno Mexicano . . .* c. 1747.

Faden, William. *A Map of South Carolina and a Part of Georgia*. London: 1780.

Fer, Nicholas de. *La Californie ou Nouvelle Caroline . . .* Paris: 1720.

Forlani, Paolo. *Vniversale Descrittione di Tvtta la Terra Conoscivta Fin Qvi*. Venice: 1565.

Franquelin, Jean Baptiste Louis. *Carte de l'Amerique Septentrionale . . .* Paris: 1909–10 [first edition 1688].

——. *Carte Gnlle de la France Septentrionalle*. Paris?: c. 1682. [Map shown is an undated later English copy entitled *The Mississippi*.]

Genoese World Map. Genoa: 1457.

Gunn, Otis B. *Gunn's New Map of Kansas and the Gold Mines . . .* Lawrence, KS: 1867 [first edition Pittsburgh: 1859].

Holmes, William H. *The Grand Cañon at the Foot of the Toroweap Looking East,* from *Tertiary History of the Grand Cañon District, with Atlas . . . ,* by Clarence E. Dutton. Washington, D.C.: U.S. Government Printing Office.

Holt, George L., Frank Bond, Fred Bond. *Holt's New Map of Wyoming . . .* New York: G. W. & C. B. Colton & Co.: 1883.

Homann heirs. *Gvinea Propia . . .* Nuremberg: 1743.

Homann, Johann Baptist. *Generalis Totius Imperii Moscovitici*. Nuremberg: c. 1704.

Hondius, Hendrik. *Accuratissma Brasiliae Tabula*. Amsterdam: 1630.

Hondius, Jodocus. *Africae Nova Tabula*. Amsterdam: 1640.

——. *Vera Totius Expeditionis Nauticae: Descriptio D. Franc. Draci . . .* Amsterdam?: c. 1595.

Jansson, Jan. *Tabula Amemographica Seu Pyxis Nautica*. Amsterdam: c. 1650.

Joutel, Henri. *A New Map of the Country of Louisiana . . . and of ye River Missisipi in North America discou'd by Monsr. de la Salle . . .* London: 1713.

Linschoten, Jan Huygen van. *Typus Orarum Maritimarum Guineae, Manicongo & Angolae . . .* Amsterdam: 1596.

Longchamps, Sebastian G. de and Jean Janvier. *L'Amerique Divisee en tous ses Pays et Etats*. Paris: 1754.

Magnus, Olaus. *Carta Marina*. Venice: 1539.

Maverick, Samuel. *Map of the State of New York*. Hartford, CT: Andrus & Judd, 1833.

Moll, Herman. *A Map of the West Indies or the Islands of America in the North Sea . . .* London: 1715? [London and Dublin: 1715–54?]

——. *A New and Exact Map of the Dominions of the King of Great Britain on ye Continent of North America* [Beaver Map] *. . .* London: 1715 [London and Dublin: 1715–54?]

——. *A New & Correct Map of the Whole World*. London: 1719.

——. *North America* [Codfish Map]. London: c. 1717 [London and Dublin: c. 1717–54?]

Münster, Sebastian. *Die Lander Asie nach irer gelegenheit bis in Indiam warden in diser Tafel verzeichnet*. Basel: 1550.

——. *Die neuwen Inseln, so hinder Hispanien gegen Orient bey dem land India ligen*. Basel: 1550.

Ortelius, Abraham. *Theatrum Orbis Terrarum* [atlas]. Antwerp and other European cities, 1570–c. 1644.

Theatrum, Africae Tabvla Nova. Antwerp: 1570.

Theatrum, Americae Sive Novi Orbis, Nova Descriptio. Antwerp: 1570.

Theatrum, Barbariae et Biledvlgerid, Nova Descriptio. Antwerp: 1570.

Theatrum, Islandia. Antwerp: 1590 [1570].

Theatrum, Maris Pacifici. Antwerp: 1589 [1570].

Theatrum, Presbiteri Johannis, Sive, Abissiniorvm Imperii Descriptio. Antwerp: 1575 [1570].

Theatrum, Tartariae Sive Magni Chami Regni tÿpus. Antwerp: 1603 [1570].

Theatrum, title page. Antwerp: 1570.

Phelps, Humphrey. *Phelps's National Map of the United States . . .* New York: 1849.

Popple, Henry. *A Map of the British Empire in North America with the French and Spanish Settlements Thereto*. Amsterdam: c. 1741 [first edition London: 1733].

Portolan chart of the Mediterranean Sea. Genoa: c. 1320–50

Pownall, Thomas. *A New Map of North America, with the West India Islands . . .* London: 1786.

Ruger, A. *Bird's Eye View of the City of Montana, Boone Co., Iowa 1868*. Chicago: Chicago Lithographing Co., 1868.

Schenk, Peter. *Novi Belgii Novaeque Angliae nec non Partis Virginiae Tabula . . .* Amsterdam: 1685.

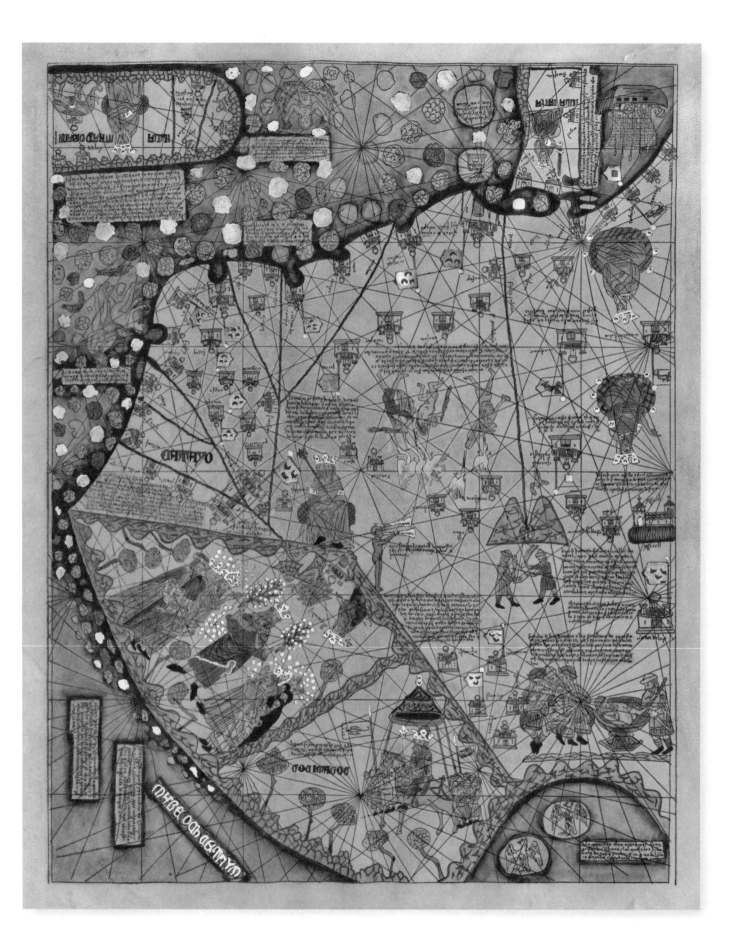

Segura, Juan. *Carte Geografica Gral de Reyno N. E. . . . .* Guanajuato, Mexico: 1803.

Schedel, Hartmann. *Das ander alter der Werlt*. Nuremberg: 1493.

Smith, John. *Virginia—Discovered and Discribed by Captayn John Smith*. London: 1624 [first edition Oxford: 1612].

Smith, L. *Planisphere*. Paris: L. E. Desbuissons and J. Migeon, 1892.

Sneden, Robert Knox. *The Battle of Gettysburg Penna. Showing Positions Held July 2nd, 1863*. c. 1863–65.

Specht, Caspar. *Tafel in Welke Vertoont Warden alle Werk-tuygen Behorende tot fe Krygs Kunde . . .* Utrecht: Anna Beeck, 1703.

Sophianos, Nikolaos. *Nicolai Gerbelij in Descriptionem Graeciae Sophiani,* from *Descriptio Nova Totivs Graeciae per vm Sophianvm*. Basel: 1545.

Sype, Nicola van. *La Herdike Enterprinse Faict par le Signeur Draeck D' Avoir Cirquit la Terre*. Antwerp?: c. 1581.

Tallis, John, Jr. *Egypt and Arabia Petraea*. London: 1851.

——. *Nova Scotia and Newfoundland*. London: 1851.

Vallard Atlas. Dieppe: c. 1547.

> *Vallard,* Aegean Sea (chart 15).

> *Vallard,* Arabian Sea, Red Sea, and Persian Gulf (chart 4).

> *Vallard,* Terra Java (chart 1).

> *Vallard,* La Java (chart 2).

> *Vallard,* Terra Java (chart 3).

> *Vallard,* North America, East Coast (chart 9).

> *Vallard,* Northwest Africa (chart 7).

> *Vallard,* Southern Africa and Southwest Indian Ocean (chart 5).

> *Vallard,* West Indies, Mexico, Central America, Northern South America (chart 10).

Vélez y Escalante, Antonio. *Derroterro Hecho por Antonio Vélez y Escalante . . . 1777*.

Vinckeboons, Joan. *Havana op 't Eyland Cuba*. Amsterdam?: c. 1639.

Visscher, Nicolaes (Claes Janszoon). *Leo Belgicus,* after Jan van Doetecum, 1598. Amsterdam: 1650.

Visscher, Nicolaes. *Orbis Terrarum Nova et Accuratissima Tabula*. Amsterdam: c. 1690.

Waldseemüller, Martin. *Tabula Terra Nova*. Strasbourg: 1522 [first edition 1513].

——. *Universalis Cosmographia Secundum Ptholomaei Traditionem et Americi Vespucii Alioru[m]que Lustrationes*. Saint-Dié-des-Vosges, France: 1507.

Wallis, John, Sr. *The United States of America Laid Down from the Best Authorities agreeable to the Peace of 1783*. London: 1783.

Worret, C. *The Siege of Yorktown, April 1862*. Washington, D.C. and Old Point Comfort, VA: C. Bohn, c. 1862

Ysarti, Antonio. *Provincia d[e] S. Diego de Mexico en la Nueba España . . .* Mexico: 1682?

Zatta, Antonio. *Il Canada, le Colonie Inglesi con la Luigiana, e Florida . . .* Venice: 1778.

——. *Le Isole Bermude*. Venice: 1778.

——. *Nuove Scoperte de' Russi al Nord del Mare del Sud sí nell' Asia, che nell' America*. Venice: 1776.

ATLAS CATALÁN, PANEL 6 [CHINA].
Abraham Cresques, Barcelona, 1375.

Akerman, James R., and Robert W. Karrow, Jr. (eds.). *Maps: Finding Our Place in the World.* Chicago and London: University of Chicago Press, 2007.

Baynton-Williams, Ashley and Miles. *New Worlds: Maps from the Age of Discovery.* London: Quercus, 2006.

Bergþórsson, Páll. *The Wineland Millennium: Saga and Evidence.* Reykjavik: Millennium Commission of Iceland, 2000.

Berkhofer, Robert F., Jr., *The White Man's Indian: Images of the American Indian from Columbus to the Present.* New York: Knopf, 1978.

Broecke, Marcel van den. *Ortelius Atlas Maps: An Illustrated Guide.* Utrecht, Netherlands: HES Publishers, 1996.

Broecke, Marcel van den, Peter van der Krogt, and Peter Meurer (eds.). *Abraham Ortelius and the First Atlas: Essays Commemorating the Quadricentennial of His Death, 1598–1998.* Utrecht, Netherlands: HES Publishers, 1998.

Burden, Philip D. *The Mapping of North America: A List of Printed Maps, 1511–1700.* 2 vols. Rickmansworth, UK: Raleigh, 1996–2007.

Cabeza de Vaca, Alvar Nuñez. *Castaways: The Narrative of Alvar Nuñez Cabeza de Vaca.* Ed. by Enrique Pupo-Walker. Berkeley and Los Angeles: University of California Press, 1993.

Campbell, Tony. *Early Maps.* New York: Abbeville, 1981.

Dudley, Edward, and Maximillian E. Novak (eds.). *The Wild Man Within: An Image in Western Thought from the Renaissance to Romanticism.* Pittsburgh, PA: University of Pittsburgh Press, 1972.

Ehrenberg, Ralph E. *Mapping the World: An Illustrated History of Cartography.* Washington, D.C.: National Geographic Society, 2006.

Fordham, Sir Herbert George. *Studies in Carto-Bibliography, British and French, and the Bibliography of Intineraries and Road-Books.* Oxford: Oxford University Press, 1914.

Friedman, John Block. *The Monstrous Races in Medieval Art and Thought.* Cambridge, MA, and London: Harvard University Press, 1981.

Gaudio, Michael. *Engraving the Savage: The New World and Techniques of Civilization.* Minneapolis and London: University of Minnesota Press, 2008.

Goodwin, Katherine R., and Dennis Reinhartz. *Tabula Terra Nova.* Dallas, TX: Somesuch Press, 1992.

Greenhood, David. *Mapping.* Chicago and London: University of Chicago Press, 1964.

Heijden, H.A.M. van der. *Leo Belgicus: An Illustrated and Annotated Carto-bibliography.* Alphen aan den Rijn, Netherlands: Canaletto, 1990.

Hill, Gillian. *Cartographical Curiosities.* London: British Library, 1978.

Jones, Gwynn. *The Norse Atlantic Sagas: Being the Norse Voyages of Discovery and Settlement to Iceland, Greenland, and North America.* Second ed. Oxford and New York: Oxford University Press, 1986 (1964).

Lemmon, Alfred E., John T. Magill, and Jason R. Wiese (eds.). *Charting Louisiana: Five Hundred Years of Maps.* New Orleans: Historic New Orleans Collection, 2003.

Lester, Toby. *The Fourth Part of the World: The Race to the Ends of the Earth, and the Epic Story of the Map That Gave America Its Name.* New York: Free Press, 2009.

Lisk, Jill. *The Struggle for Supremacy in the Baltic: 1600–1725.* New York: Funk and Wagnalls, 1967.

Martin, James C., and Robert Sydney Martin. *Maps of Texas and the Southwest, 1513–1900.* Austin: Texas State Historical Association, 1990 (1984).

Reinhardt, Stephen G., and Dennis Reinhartz (eds.). *Transatlantic History.* College Station: Texas A&M University Press, 2006.

Reinhartz, Dennis. *The Cartographer and the Literati: Herman Moll and His Intellectual Circle.* Lewiston, NY/Lampeter, Wales: Edwin Mellen Press, 1997.

Reinhartz, Dennis, and Charles C. Colley (eds.). *The Mapping of the American Southwest.* College Station: Texas A&M University Press, 1987.

Richardson, William A. R. *Was Australia Charted Before 1606?: The Java la Grande Inscriptions.* Canberra: National Library of Australia, 2006.

Schwartz, Seymour I., and Ralph E. Ehrenberg. *The Mapping of America.* New York: Abrams, 1980.

Southworth, Michael and Susan. *Maps: A Visual Survey and Design Guide.* Boston: Little, Brown, 1982.

Temple, Robert. *The Genius of China: 3,000 Years of Science, Discovery, and Invention.* London: Prion, 1999 (1986).

Thrower, Norman J. W. *Maps and Civilization: Cartography in Culture and Society.* Third ed. Chicago and London: University of Chicago Press, 2007 (1972).

Trickett, Peter. *Beyond Capricorn: How Portuguese Adventures Secretly Discovered and Mapped Australia and New Zealand 250 Years Before Captain Cook.* Bowden, S.A., Australia: East Street Publications, 2007.

Tufte, Edward R. B*eautiful Evidence.* Cheshire, CT: Graphics Press, 2006.

——. *Envisioning Information.* Cheshire, CT: Graphics Press, 1990.

——. *The Visual Display of Quantitative Information.* Cheshire, CT: Graphics Press, 1983.

——. *Visual Explanations.* Cheshire, CT: Graphics Press, 1997.

Vásquez, Carlos, and Verónica Iglesias Swanson. *Nao de China: The Manila Galleon Trade, 1656–1815: An Historical Exhibit.* Albuquerque, NM: National Hispanic Cultural Center, 2009.

Weber, David. *Bárbaros: Spaniards and Their Savages in the Age of the Enlightenment.* New Haven, CT: Yale University Press, 2005.

Woodward, David (ed.). *Art and Cartography: Six Historical Essays.* Chicago and London: University of Chicago Press, 1987.

Woodward, David, J. B. Harley, and G. Malcom Lewis (eds.). *The History of Cartography.* 3 vols. Chicago and London: University of Chicago Press, 1987–2007.

THÉÂTRE DE LA PAIX ENTRE LES CHRÉSTIENS ET LES TURCS. Anna Beeck, from *A Collection of Plans of Fortifications and Battles,* The Hague, 1684–1709.

PICTURE CREDITS

**FAIRWINDS ANTIQUE MAPS,
NEW YORK / FAIRMAPS.COM**
91: Zatta, Antonio. *Le Isole Bermude*. Venice: 1778.

GOOGLE BOOKS
56: *De Insulis Nuper Inventis* (1493); 135: *Brevisima Relación de la Desctucción de las Indias* (1552)

COURTESY OF THE HUNTINGTON LIBRARY, SAN MARINO, CA. WE ARE GRATEFUL FOR THE PERMISSION OF M. MOLEIRO EDITOR TO REPRODUCE THE VALLARD ATLAS MAPS; SEE FACSIMILE EDITION AT WWW.MOLEIRO.COM
VALLARD ATLAS/http://sunsite3.berkeley.edu/hehweb/HM29.html:
4, 58t: West Indies, Mexico, Central America, Northern South America (chart 10).
57, 139, 146: Northwest Africa (chart 7)
58b: Aegean sea (chart 15)
70–71, 137: Terra Java (chart 1)
72: Terra Java (chart 3)
74: Vallard Atlas, La Java (chart 2)
109: Southern Africa and Southwest Indian Ocean (chart 5).
138: Arabian Sea, Red Sea, and Persian Gulf (chart 4).
149: North America, East Coast (chart 9).

JOE LEMONNIER / MAPARTIST.COM
xvi: Illustrated Africa map

COURTESY OF GEOGRAPHY AND MAP DIVISION, LIBRARY OF CONGRESS
ii–iii g3190m gct00059; v g3200m gct00003; vi–vii, 5, 77, 108: g5672m ct000821; xii: g3200 mf000001; xiv–xv, xvii, xviii, xix, xx, xxi, 14–15, 114–15, 158: g3300 ct000232; xxii, 6–7, 25tr, 93, 150tl, 152tl: g3880 ct000377; 8: g4411a ct000023; 9, 20–21, 22, 23t, 38: g4390 ct002323; 10–11: g3200m gct00001; 12–13, 80: g3200 mf000070; 16–17, 28t, 112–13, 152br: g3715 ct000001; 18–19, 28b: g3300 ct000668; 24, 25tl, 59: g5780 ct000403; 26–27: g3200 ct001471; 29: g4030 ct000530; 32–35, 162: g3200 ct000725; 36–37: g3200m gct00035; 39, 60–61, 151t: g3300 np000058; 40, 48–49: g3201s rb000011; 41: g4924h lh000348; 43, 148t: g3290 hl000010; 44, 102–3, 159l: g3700 ct000761; 50–51: g3201s ct000130; 52–53, 97: g3200m gct00003; 54–55, 83: g3290 ct000342; 62: g3884y ar146200; 63: g3861a cw0001700; 64–65: g3884y cw0673700; 67: g3200 ct001473; 73: g3200m gct00087; 75: g3200m gct00003; 77: g3200m gct00003; 78–79, 143: g8200 ct001455; 82: g6810 ct001164; 87: g8200 ct002076; 90: g3201cm gcws0183; 94–95: g7000 ct000625; 99, 136b: g8735 ct000313; 100–101, 136t: g7823g ct000750; 108: g3200 ct002087; 111: g4042m ct000784; 114–115 [background]: g4332gm gnp00002; 116–17: g4410 ct001000; 119, 120–21: g3300 ct003436; 124, 187b: g4260 ct001852; 125: g4154m pm002230; 130, 172–73: g5730 ct000158; 131: g3200m gct00035; 140: g3200m gct00003; 141t: 007.00.05; 144: g3910 ct001281; 145, 147: g5400 br000098; 151b: g4300 ct001515; 153: g4410 lh000552; 155: g3300 ar002700; 154–54 [background]: g4390 ar168800; 160: g3200m gcw0013960; 163: g5700m gct00127c; 164: g5700m gct00127c; 166: g3700 ct000080; 167: g3700 ct000760; 170–71b: g3201cm gcws0183; 171: g3201cm gcws0183; 177: g3804n pm006021; 179: gvhs01 vhs00264; 180–81: g5700m gct00127a; 183: g8300 ct000421; 185, 186–87: g3800 ct002319; 188–89, 190: g3804n pm006011; 192–93: g5700m gct00127a; 197: g3200m gct00001; 202: g3200m gct00215; 207: g5700m gct00127c; 208: g3200 ct002087.

COURTESY OF PRINTS AND PHOTOGRAPHS DIVISION, LIBRARY OF CONGRESS
3b: LC-USZ62-110320; 134: LC-USZ62-3032; 150br: LC-USZ62-53338; 159r: LC-USZ62-671; 184: LC-USZ62-99650

COURTESY RARE BOOK AND SPECIAL COLLECTIONS DIVISION, LIBRARY OF CONGRESS
3, 56, 74 [background]; 84: 117.00.00; 85: 116.00.00; 86: 115.00.00

BARRY LAWRENCE RUDERMAN ANTIQUE MAPS INC., / RAREMAPS.COM
viii–ix, x–xi Huygen van Linschoten, Jan. *Typus Orarum Maritimarum Guineae . . .* 23b, 96, 113t: Châtelain, Henri-Abraham. *Carte Tres Curieuse de la Mer du Sud . . .* 45, 122–23: Tallis, *Nova Scotia and Newfoundland*.
81: Münster, Sebastian. *Die Lander Asie nach irer gelegenheit bis in Indiam warden . . .* 107: Ortelius. *Theatrum, Presbiteri Johannis, Sive, Abissiniorvm Imperii Descriptio.* 126: Bunting, Heinrich. *Asia Secunda Pars Terrae in Forma Pegasi.*

133: Schedel, Hartmann. *Das ander alter der Werlt.*
165: Longchamps, Sebastian G. de and Jean Janvier. *L' Amerique Divisee en tous ses Pays et Etats.*

DAVID RUMSEY MAP COLLECTION, CARTOGRAPHY ASSOCIATES / DAVIDRUMSEY.COM
105: Zatta, Antonio. *Il Canada, le Colonie Inglesi con la Luigiana, e Florida . . .* 157: Delisle, Guillaume. *Carte du Canada ou de la Nouvelle France . . .* 161t, 161bl: Gunn, Otis B. *Gunn's New Map of Kansas and the Gold Mines . . .* 168–69: Berghaus, Heinrich. *Geographische Verbreitung der Menschen-Rassen.*
176: L. Smith. *Planisphere*. Paris: L. E. Desbuissons and J. Migeon.
191: Thomas Jefferys advertisement.

ANTIQUARIAAT SANDERUS, GHENT, BELGIUM / SANDERUSMAPS.COM
1: Jansson, Jan. *Tabula Amemographica Seu Pyxis Nautica*. Amsterdam: c. 1650.

SHUTTERSTOCK
3t: © Shutterstock/ribeiroantonio

UNIVERSITY OF TEXAS, ARLINGTON, SPECIAL COLLECTIONS
30–31, 110, 148b: Waldseemüller, Martin. *Tabula Terra Nova.*
98: Segura, Juan. *Carte Geografica Gral de Reyno N. E.*
156: Joutel, Henri. *A New Map of the Country of Louisiana . . .*

WICHITA STATE UNIVERSITY LIBRARIES, DEPARTMENT OF SPECIAL COLLECTIONS
161br: Gunn, Otis B. *Gunn's New Map of Kansas and the Gold Mines . . .*

WIKIMEDIA COMMONS
46–47: Ortelius - Maris Pacifici 1589; 66, 76: Island 1590 Theatrum Orbis Terrarum Ortelius; 68, 88: Carta Marina; 104, 127: 1583 Leo Belgicus Hogenberg; 118: 1776 Zatta Map of California and the Western Parts of North America - Geographicus - AmericaWest-zatta-1776/Geographicus Rare Antique Maps; 128–29: 1650 Leo Visscher; 141b: Braun Cairo HAAB; 142l: Braun Mexico UBHD; 142r: Braun Cusco HAAB; 174–75: Imperial Federation, Map of the World Showing the Extent of the British Empire in 1866/http://maps.bpl.org/Fae.

WORLD

MERCATOR'S PROJECTION

Published by
JOHNSON AND WARD